illennial
Age of Chivalry

Brice Parker

CLAY BRIDGES
PRESS

In loving dedication to my mother,
Kimberly,
for raising me to be a gentleman.

TABLE OF CONTENTS

Introduction   1

Chapter One: The Millennial Perception   **5**
    A Millennial Introduction   5
    Social Constructs   13
    Connectivity Complex   20
    Media Energy Scale   27
    Chivalrous Morality   33

Chapter Two: The Millennial World   **39**
    Sociological Relativity   39
    Netiquette and Techlination   45
    Millennial Timeline   53
    Dating/Hookup Culture   60

Chapter Three: The Millennial Gentleman   **69**
    Chivalry—Then and Now   69
    Millennial Knights   76
    Five Noble Vestiges   84
    Nobility to Majesty   92

Chapter Four: The Millennial Relations   **97**
    Contemporary Love   97
    The Three Ps   104
    Millennial Disposition   109
    Monogamy Mechanics   116
    Pay It Forward   121

Chapter Five: The Millennial Message   **125**
    The Millennial Perception   125
    The Millennial World   133
    The Millennial Gentleman   139
    The Millennial Relations   146
    The Message   153

Notes   **163**

Introduction

The following generational memoir addresses the millennial age of chivalry. Immediately, this may seem like an oxymoron as it's almost unanimously agreed upon that chivalry is dead. Millennials are seen as both cause and effect of this, but our relational ADD is only the cherry on top of a dysfunctional generation sundae. Despite the amazing things to come from our time, we've been categorized as a generation with some pretty bad labels. Some key examples include being entitled, impatient, stubborn, unfocused, and being called snowflakes. The unsettling millennial reputation under speculation is concerned with egocentricity, and this actually ties in with the previous traits as a cornerstone to our relational behavior. This is where chivalry steps in.

I was born and raised in Los Angeles, exposing me to a sort of "Hollywood state of mind" with regard to a higher degree of superficial and materialistic matters. Celebrity culture has always been a part of our history, but has been vastly inflated during this millennium by the increase in media. As time went on, I learned about alternative lifestyles abroad in a polar opposite romance culture of Paris, France. A culture shock experience gave me an appreciation for things not as common in Western culture and a different capacity for the relevance of media to my life. In hindsight, seeing the way other parts of the world interacted in their relations, it is no wonder why I had such ambiguity with media in mine.

Millennials were the first generation to inherit the internet, which single handedly took hold of our world with a magnetism to social and relational media. It was as if we were the test subjects, but also the creators. We used network technology among the Silicon Valleys to industrialize Western culture with a new media consumer-based marketplace. As time went on, this became a formidable part of our everyday lives, and innovative utilities began to shape our general societies and our personal psyches. The effects of these systems on relations mainly stem from the objectification of subjective matters in the dichotomy of their coexistence online. This being said, the real world to media translation is not inherently good or bad; its value lies solely in our own discretion of application.

This timeline is a condensed glimpse of the event horizon surrounding the millennial and only scratches the surface of why we have become what we are. The idea of chivalry has been around for hundreds of years, but it would seem that they would now be of the utmost necessity due to its apparent demand. With the coming of the extensive newfound outlets of connectivity, there has been widespread potential for almost endless interaction in forming and conducting a specific agenda. These media have the same amount of potential to be prolific to our relations and detrimental to our culture, and vice versa. It is imperative that we learn to properly utilize these things and remain disciplined in our diplomacy.

As isolated as media matters would appear to be to their own realm, the effects are becoming more prevalent in our immediate environments. There is disconnect in so many outlets that are actually put in place for just the opposite: flat screens that cover restaurant walls, texting in the middle of in-person conversations, and taking pictures to share the moment with those not there. The ever-changing nuances of our behavior are minimal compared to the blatant misuse on the back end that we often don't see, where internet bullies, trolls, or presidents of a free country continue to abuse these platforms. Through chivalry, many of the challenges we face today can be solved, and it entails only a minor shift in perception and the actions that follow.

In putting the ambition for a utopian world aside, we see the option of dealing with a more manageable task in bettering ourselves. Once we begin to conduct ourselves in an honorable manner, we will receive the bounty of positive relational implications. The most effectual aspect of chivalry in our millennium is the switch in mentality that involves everyone in its practice. Chivalry is about the personal decisions to act in a manner that serves your community, and since by definition this includes everyone, it requires that everyone participate. Each one of us has the capacity to practice chivalry and obtain the benefits that it has to offer in the unique facets of their respective lifestyles. We need to become more progressive in our perceptive equality for this to materialize.

Media was developed to bring us together, but since we were all new to their devices it wasn't always seamless in facilitaticn. Chivalry is the means by which this universal connection can be properly carried out in bringing all walks of life into loving fruition. Chivalry is the answer to challenges that expand far beyond social boundaries into the literal preservation of our planet in a home we can share. Understanding is the basis of civility, and the treatment of one another will come full circle to serve the greater good of civilization. It all starts with the individual, and you are the most beautiful when you are the most yourself. These media connect us in a universality, but we must retain our unique singularity and not excessively follow others.

Millennials have been through the good and bad that comes with inheriting this advanced media, so it is our responsibility to learn from this anomaly in looking to the future. The next generations will not know what it is like to grow up without these things, so it is imperative that we supplement more natural stimulation in retaining their moral integrity. At the end of the day, love is strength, which is why we need to be like knights in bravery and unyielding vigor toward the challenges that face our future. If you are given the ability to do the right thing, you have a responsibility to do it. Whether it should be alongside your brothers and sisters or alone, you must harness your strength and choose love.

Chapter One

The Millennial Perception

A Millennial Introduction

The influence of millennial connectivity and subsequent media has reached numerous aspects of our relations, and by extension, the concept of chivalry. The millennial generation was exposed to the dawn of the twenty-first century, fostering us with unprecedented growth in global integration, and as such the entire dynamic of our relational behavior was changed. The invention of the internet is nothing short of a pivotal historic milestone that has changed the world in more ways than we can measure. Not only were we brought up by this uncharted and unlimited access to information and one another, but we were the first to be. As a result of the ability to access connectivity that continues to expand and evolve at such a young age, we have learned to master and pioneer all of its components, but still, unforeseen effects ensued. Comparatively speaking, the generations prior to the millennium could hardly fathom the capabilities that we've become so accustomed to, so it is important to bring them up to speed, and even more importantly providing insight for future generations. Being familiar with these new abilities can be prolific by its inherent potential, but with great power comes great responsibility. With this power, it is imperative that we avoid adopting a laissez-faire attitude toward these things, because by nature, a careless mindset opens the way for negligence and consequences unseen even by its own benefactors. To properly understand the importance of our actions and our legacy, we need to first have perspective on just how influential these current times are. In the past 100 years, the rate of growth in technological industries has been incomparable, and although some of these breakthroughs are undoubtedly for the greater good of humanity, it is not the technology itself that creates or destroys, but rather how we use it. With respect to chivalry, this relatively recent influx has brought new elements to the entire world population in how we conduct our relational behavior, and by extension, how we treat our loved ones. Metaphorically comparative to the discovery of fire, the invention of the internet serves to be a potential stepping-stone for humankind and what we are able to create as a product of its ingenuity. Consequently, this also has a potential of spreading quickly and out of control, causing destruction. Ultimately, we have control over the utility of our mechanics, and confident as we are with these novel capabilities, the effects can be precarious and have been proven to be of a high significance with the ways we interact and relate with one another. These capabilities can be applied to many types of relations in many useful ways; however, it seems that certain aspects have proven to be critical to the more sensitive personal relationships. Communication is said to be the key to any successful relationship, but the influx of connectivity has changed the very nature of communication as we know it. It has changed the grand scheme of how we perceive relations starting with how we conduct day-to-day acts with our loved ones. This media has taken the role of being the new standard of relational measurement in how we gauge who we are involved with. It is undeniable that the connectivity outlets brought with the passing of the twentieth century have established a solid footing in human interaction, but instead of pointing out which ones are somehow inherently right or wrong, rather observe the ways in which we use and abuse them. As relationships are undoubtedly the most intimate and fragile form of human interaction, it is no wonder why there has been turbulence in this field lately. However, with regards to chivalry, a steady sea never made a skilled sailor. So, instead of grieving that we had the burden of establishing precedent and learning the hard way, try to see our time spent as earning an understanding of both ends of the spectrum, of its existence or lack thereof, and everything in between its conception to current and future times. Using this model, we can then observe culture dating from medieval times when chivalry was at its peak, to modern times and where we may find ourselves in the future. This wider range of perception will serve to help you gain a dynamic and multifaceted understanding to enhance your utilization and discipline. With a position such as this, it is our moral and ethical duty, to those we love and future generations, to lead by example in a new era of capabilities facilitating a better world.

My personal understanding of relational sociology was initially shaped by Western culture, but expanded to the perception that we are moving toward globalization. In 1990, the World Wide Web was introduced through the internet, and shortly thereafter, media took form in Western society by the godfather of the internet, AOL (America Online). The lateral equivalent of this internet empire is the globalized version of Google, showing connectivity and relational media expansion alongside me and my Western millennial generation. I was born in 1992, and raised in Los Angeles, California, in what seemed to be the perfect recipe for inheritance of a full effect in part of the millennial zeitgeist. With its standing as the final frontier in Western civilization, Los Angeles has led the way in entertainment and media embodied by the spirit of the world-famous Hollywood. LA's cultural background along with various trivia facts giving way to the current vibe allows this city to hold a place in contention as being one of the major metropolitan subcultures of the world. A first-world bubble with emphasis on media importance in a time of newfound connectivity can open way for light to be shed on the more superficial and sociopathic attitudes toward relations. Luckily for me, I was afforded the opportunity of separation to cultures much more rooted in traditional heritage as I took part in several different study-abroad opportunities throughout college. This allowed me to experience firsthand the contrast of certain westernized subcultures to that of vastly different international students based around the world. My studies ranged from grandiose cities such as Paris to the likes of more bosomed towns along the Swiss Alps, providing a microscale of contrast in culture. These travels served to later reveal themselves as much more than scholastic endeavors, serving as inspiration among one of the most prominent cultures to embody food, art, and love: French. This has provided me with an illumination to the bounty attained by genuine and meaningfully maintained relationships. I began to find a part of myself while somewhat lost along the chateaux-scattered countryside flourishing with some of the world's best grapes. However picture-perfect these ancient castles, pastoral towns, and vintage wineries were, what I found to really inspire me were moments with the ones experiencing this with me. Even in the midst of climbing the Eiffel Tower or walking the streets during the La Fête des Lumières, I can most clearly remember the looks on people's faces and the shared feeling of those times. This being said, there may be an overture of *hommages française* throughout the following content, but there is certainly a strong correlation between their culture and chivalry. As a matter of fact, the origin of the word *chivalry* is literally derived from the French word *chevalerie,* denoting a person who is master of the horse; the term was later used to refer directly to the knight. Along with the more obvious parallels, you can note more microanalytical points of French culture, such as how the French language still utilizes formalized verb conjugations in addressing one another. There is empirical proof supporting the correlation of their culture and the very idea of chivalry, but the abstract concepts are more prevalent in finer details. One of these subtleties is engrained in the idea of tradition in that throughout European culture, people strive to maintain older, yet timeless, values which pertain to chivalry such as the family name, communal ethics, and so on. This and many more indefinable cultural dispositions of Europe prove to embody these arts and must be experienced to be understood. Fortunately, so much has been preserved for intents and purposes associated with the art of relations, offering those who wish to learn the accommodation of a vicarious feeling when studying its history. Given the timing of these studies in conjunction with global connectivity, it is unavoidable to see contrast in the world outside Western society as well as our social connectivity—particularly in the ways of chivalry. This book serves to be my *pièce de résistance*.

Chivalry is a subjective concept, meaning it is subject to interpretation that varies throughout different cultural norms and the times that we find ourselves in. As with anything subjective, there is a degree of personal philosophy, so contemplation and an open mind will advance your competence. There's a wide variety of philosophical views, but early schools practiced around the Middle Ages and the conception of chivalry seem to be more applicable, like those earlier proposed by Aristotle, one of my favorite philosophers. Although chivalry is the topic at hand, the ramifications upon everyday relations and socialization expand far beyond those involving love and chivalry. Chivalry is a practice of serving the greater good and having honor and dignity in all that you do with elevated meaning. The word *chivalry* implies a code of characteristics most commonly noted by a gentleman, but it means much more than this or the sole association with knights of medieval times fighting for glory and courting women. The purpose here is not to be mistaken with the intention of targeting males or inferring that any group is the focus. This being said, what I believe to be true about chivalry is subjective to my experience, and it is not my intention to impose it on anyone, but my hopes are that this content may be utilized by everyone. Hopefully, after sharing my point of view, you will see that chivalry transcends all borders and everyone should learn to embrace their part in its tradition. Although technically impossible in subjectivity, it is my job to be unbiased and not to speculate in my own personal regard, but rather to speak in general for you to contemplate your own perception. Perception is taking content and determining its value to circumstantial application, so content lined with insight is what I provide, and it is then your decision as to its relevance and importance in your own life. These connectivity tools have potential to assist us in countless beneficial causes in hopes to advance humanity and produce amazing rewards from accomplishments of our progress. However, it is also easy to choose counterintuitive behavior and focus on using these capabilities for indulgence in self-serving tendencies, which coincide with the demand of certain societal standards. This is not to say that people are sociopaths by nature, but rather societal pressures can promote certain addictive qualities that can be damaging to certain parts of the individual, and by extension certain constituents in society itself. Pure intentions to achieve a sense of genuine fulfillment in your life can be clandestinely derailed by the enticing prospect of instant gratification and notoriety that our current society offers. An increase in this type of mentality has come as a result of our generation and in what we supply and demand; it is highly conjunct with an economy. This can be noted through observation of certain marketing that causes consumerist behavior and materialism in the United States' Western society, fueled by capitalism. Never has there been so much value placed on information delving into the various marketing strategies that larger corporations, especially those with online interests, attain and seek out among individuals. This is seen at your first use and updates in the lengthy contractual agreements they require you to sign and accept to make a profile. Societal systems in media coinciding with economics perpetuate and support an excessive consumerism, but more specific agents affecting societal ethics like the media contribute more directly to our lifestyles and relations. Not to say that this is necessarily a full representation of our direction as humankind, but in these times, it is similar to a gold rush in the influx of social networking concentration. The effects can be found in the most basic derivative attributing to this societal behavior, in that we have become overly objective in our morals. The concern here is in how it is unquestionable that the treatment of life as an object takes away from our humanity.

Social Constructs

Much of the current disposition toward relational deficiencies is brought by imposition of objectivity fueled by certain media in our newfound connectivity. In contrasting metropolitan subcultures such as Los Angeles, Paris, London, Tokyo, and so on, to smaller cities and towns, this objectivity is more commonplace. Unlike the majority of the world, these highly concentrated areas of population emphasize the attributes and accolades that define our character with respect to quantity instead of quality. This is due to various aspects of their being shared with the products of their environment provided by these media systems and the mentality they promote. In current times, we are seeing more people in these larger cities and more people among these media systems because of a desire for more connectivity. This, mixed with the recent innovations in media, leaves chivalry susceptible to certain stunts in growth. In the millennial age, how we see ourselves, and by extension each other, has been heavily influenced by the numerical scale of value adopted from the realm of quantifiable representation. This is nothing new though, as humankind has derived value this way ever since the creation of math, but too much can lead to an unhealthy imbalance of objective mentality. A perfect example of this concept is in the value we tend to place in monetary quantification and how, although it does have a measurable value, there are things money can't buy. This trend is more prevalent than ever in how quantifiable things have become more pertinent to how we value our character. This is especially relevant with regards to the concept of social networking in how we find self-worth signified by counting comments, likes, shares, followers, friends, and so on. A friend is not an inanimate object or something to be quantified; a friend is someone who should be immeasurable in importance to our lives. The enduring effectiveness of this concept derives strength from the fact that these systems allow you to portray yourself in a light best suited to receive measurable validation. The problem here is that this validation is restricted to the confines of these systems and leads to ambivalence of attainable fulfillment between reality and your online alter ego. In producing an alter self-image, you conform to what yields the best results in a platform of representation, converting this alias to the real world. By doing this, you confide yourself to these parameters and overlook the most vital aspect of a fulfilled life, being unbound by any criteria of personal expression. In studying human behavior and relations, it is clear to see that love has nothing to do with neither measurement nor objectivity, but it is more about genuine value and feelings of intuition. Even with all of the current access to information about one another, there is still no possible way to know everything that has or is currently happening to a person; therefore, the only logical solution is a subjective approach. Some of the relational objectivity in modern media can be traced back to different measurements of importance in outdated societal gender roles. This is one of the main roots in this blossoming trend in how men and women overcomplicate the fundamentals of our role in society and personal relations. The two genders have historically coexisted maintaining very different societal roles, but we are moving toward equality. This ideology translates into our relations in what seems to be a never-ending power struggle when really we should be finding balance. With an objectified view of relationships, we often tend to oversimplify the classification of our efforts and productivity, which can result in an inaccurate breakdown referred to as "what you bring to the table." This is an unfair setup for clear perception of relations because it does not account for the many qualitative components of a relationship. This is not to be mistaken as blaming societal roles and the collective thinking of our cultures for this phenomenon. It is impossible to indict the system because it is a reflection of what we allow, so instead let's look at a microlevel in our own behavior as being those parts. By embracing ideas in an overgeneralized mentality of a system, we only perpetuate stereotypes, which are more often misleading than not. The truth is, we both share all the same attributes in different circumstances, and it is all in a case-by-case basis. All differences considered, the basic principle remains that we simply need to work with instead of against each other. The systems that emphasize the value of measurable validation exploit certain stereotypes to further perpetuate their agenda. This train of thought promotes that men are most commonly associated with being primarily right brained, using strict reason to guide their actions. Likewise, the stereotype assumes that women are primarily left brained, using pure emotion to guide their actions. For initial intents and purposes, these differences boil down to creating a false societal generalization creating bias in who we are and stops our collective progression toward transcendent chivalry for all. We must remember that we are all individuals and retain the right to act on our own accord without the guidance of these enigmas of free will that allow for anyone to act with chivalry.

Men and women are fundamentally different, but this does not justify different treatment or the imposition of different values between us, including chivalry. Historically speaking, for basic biological reasons, society has subjugated women in almost every manner of basic power resulting from primitive times when brawn actually triumphed over brain. Today, this is certainly not the case; in fact, it is the other way around, which is why we are seeing progressive mentalities in general areas surpassing relations. Although in general, we have become more sophisticated than our Neanderthal ancestors, even the new world in a westernized nation of the United States has yet to elect a female president, and a woman's voice in the voting process has only been legalized within the past 100 years. However, although times are quickly changing and there are measurable matters that are improving, we are still far from equality as a mentality among humankind. Feminist movements extend beyond a state of mind, looking for basic control over what happens with their bodies. We currently see this ranging from a more serious matter in pro-life debates to one of the more liberal movements such as "Free the nipple." As obvious as this would seem in this context, it is still a major issue throughout all cultures of the world, as some insist that certain female "indecent exposure" be punishable by death. By this standard, relatively speaking Western culture is fairly progressive, but we still have a long way to go on a global scale. The millennial generation has been presented global connectivity with the task of maintaining our direction in progressive rights and treatment of one another. It is important that we use these systems in a productive and positive manner instead of abusing them and letting them control us to perpetuate outdated ways. This being said, the systems are inanimate, so the focus lies on the individual and each person's role in millennial progression toward love and chivalry. As I have previously endorsed, I am a man and will try my best to not speak for women, but there is definitely something to be said. Although women are typically excluded from the conversation of chivalry, they have always had a place there. At this point, it is pure ignorance to overlook their involvement with this idea throughout history. Especially now with capabilities in global networks and dissemination of information, it is our duty to raise awareness of previous human shortcomings and make a change for the better. The millennial era is a pivotal time in solidifying proper dispositions toward equality. With regards to a recent depiction of a female in the media, we see this continuation of a primitive mindset, echoing the idea of their dependency on men. A general example here is even recognizable to a child, taking the form of bedtime stories and animated pictures, involving a princess and prince charming. The agenda of the typical female role here is almost always personifying the damsel in distress, desperately in need of a male savior to come to her rescue. As far as this type of overlooked recognition goes, the correct version of an appropriate endorsement of reality in the female condition comes from a predominantly feminine culture: La France. This country is a beacon of fashion, art, fine dining, and just about everything else to be considered romantic, but this is not to be mistaken with a type of weakness. The French make this concept glaringly apparent through the following endorsement of the feminine and the special title for a woman who is not only irresistible by demeanor, but surreptitiously cunning, divinely charismatic, and forthright lethal. The *femme fatale* incites the true meaning of potential power in regards to the female condition giving just a glimpse of deserved recognition. It can be seen in the tip of a spear overthrowing suppression, wielded by none other than the revered knight and freedom fighter of France, Joan of Arc. France is not the only culture to adopt such a viewpoint, as the entire world can recognize the classic endorsement, "Mother Nature," referring to matriarchal importance and the place of women in the world, as the world. Luckily, we live in a time when just within the last century, women have fought to make triumphs over several blatant injustices. This is one of the profundities embedded in these media systems, in that they provide us with a global platform to be seen and heard. Now is the time with societal standards on the cusp of transformation to push the envelope; let your voice be heard, stand up for what is right, and make a change.

The root of chivalry, gallantry, and honor is the simple idea of personal ethics, which can be tampered with through media contribution to collective thinking. Needless to say, we do not live in a perfect world of progressive systems, but the bright side is that these media currently offer a plethora of innovative outlets in individual focus, which can lead to a collective movement toward a better world. Through the power of this connectivity, a single individual can reach the entire world, making a pivotal difference to the present and to future generations, by simple means such as being a good example in these systems, your community, and personal relations. Much of this potential for change has presented itself on very large platforms in our millennial era, so it is important to not underestimate the implications of your actions. On the internet, this concept is especially relevant as literally everything entered is recorded and saved regardless of your attempts to cover up or delete your content. This goes to show the importance of personal censorship and appropriately refining what you mean, instead of folding under the pressures of what is currently cool or trending on these systems. It seems as though more and more men and women are thinking less independently and focused more on replicating what public figures tell them. Certain public figures use these systems to promote their agenda, and the systems benefit from their involvement, so it tends to perpetuate this type of collective thinking. This lack of individual identity has proven to be a major contributing factor leading to a chivalry epidemic. The disparity here can be attributed to many causes, but pertaining to the media's involvement, the main cause is money and control. The media is controlled in this manner primarily because it takes its form through only a handful of super corporations: General Electric, News-Corp, Disney, Viacom, Time Warner, and CBS. These six super corporations own rights to and control a vast majority of the media involving television, radio, and newsprint. You can be assured that at the very moment in conception of the internet, they had their hands all over that too. Filled with clever marketing tactics designed to tell us how to think and what to buy, mainstream media manufactures public opinion. The catch-22 of our generation is that even though media has been so saturated with corporate pressure, there has never been so much access to truth if you only know what to look for and how to do so. It can be easy for an individual caught up in the agenda of these systems to succumb to a set of ethics derived from what is advocated in the spotlight. The point being here is that these media do not always have your best interests in mind as an individual, and you must learn to differentiate between the two. The idea that we must look inward to find peace, and not through means of external validation, is the essence of success in chivalry.

Connectivity Complex

Relations depend on acceptance of individual identity, which can be lost among the societal collective thinking we are conditioned with through connectivity. Embracing individuality helps shed some of the weight of societal impositions and the devices that are used in partially limiting people into seeing through the filters of mainstream agenda and conformity. One way to counter these external pressures and the type of thinking they provoke is to avoid projecting in a forceful manner, and instead find a common ground where you can share feelings and compromise for the sake of progress. Understanding these mechanics will give way to ability in effectively utilizing the potential in all the various systems we have, to facilitate the experiences and relations you desire. All things considered, there has never been a better time to see and display individual prerogative with the ongoing acceleration of global connectivity. Although there may be lots of interference in shaping our perception of relations over media, this does not mean chivalry has no place in these systems. As much influence as these systems seem to impose, they offer an equal amount of opportunity to stimulate creative ideas in finding individual truth. The truth is, you will find that humankind is just that—kind by human nature. This is the basis of chivalry in its simplest form: Nice is not just a place in France. No matter how caught up we are in the allure of things that are not what they seem, there is always room for happiness and love if you open your heart to them. You must learn to accept people as they are and not what you think of them because you will never fully understand anyone or anything; everyone has secrets, and what's more, things that they don't even know about themselves. This makes it very difficult to truly relate to one another, let alone everyone else, especially when you base this perception off of something as one-dimensional as a profile page. This and the sheer number of people out there draw a wide spectrum of the type of relations you are able to form. The key is in understanding yourself, and that although we are all different, we are more alike. If we can change our perception at this level, movement toward relating to one another can be in our individual control leading to a global shift of collective acceptance. In a time of increasing emphasis on global social influence through media, it is important to project a true persona, thus attracting your true relations. With the ever-increasing capabilities and freedom of expression inherently involved with the internet, we have a degree of control over our social surroundings that allows us to choose exactly how we project ourselves to others. This does not necessarily suggest creation in projecting a reality around you tailored to your imagination, but points toward internal means and what you can do within your perceived limitations. In this regard, you have complete and total power over your body, mind, and soul; therefore, you should not let anything or anyone control you. As such, you also must accept that you do not have that power over others. The power you have over others is the ability to choose whom to surround yourself with, especially in this day and age where a majority of your input stimulus surrounding influences are connected by clicking "add friend/unfriend" or "follow/unfollow." What matters here is not how you can affect the world around you, but rather how you let it affect you, and how you synergize the functions to your relational disposition.

Comparing today's societal standards in relationships to those of earlier times, it seems that this media influence has facilitated outlets for short-term relations. It's almost as if these systems have promoted our relations to become more disposable and less important. Of course, this statement is tiptoeing the line of attributing correlation to causation, but there is no such thing as mere coincidence. There has to be some significance to this incurrence of these systems and the hard statistics of divorce rates being at an all-time high. Dysfunctional relationships have always been around, but nowadays the reasoning behind our drama and attempts to amend our troubles is pathetic. On the contrary to today's deficiencies here, there was a time brought to light by chivalry and the poetic interpretations of William Shakespeare depicted in tales such as *Romeo and Juliet*, or *Antony and Cleopatra*. Other epics including the likes of Achilles and the Trojans, conceived on the basis of a love affair and the legendary war that ensued by the grace of Helen. This fabled conflict in the love triangle even arose between the golden standard of knighthood between Sir Lancelot, King Arthur, and Guinevere. It seems that in these far past times, there tended to be more significance to the failure or success of these relationships. Loyalty in friendship, family blood feud, and even warring nations all exemplify how we find reasons comparatively petty and disproportionate in modern times. Today, relationships end because someone liked the wrong picture. Back in medieval times, we would have heard tales of chivalry glorifying ideas such as true love and even the ultimate sacrifice of one's own life for the chance to elope with one's lover. Diverging away from this era as time goes on, it seems more and more that all we care about are celebrity hookups and who the current reality show bachelor/bachelorette will choose. Instead of the face that launched a thousand ships, we have the ass that launched a thousand memes and broke the internet. We could easily by focusing on better content and functions to strengthen our relations, but we are just too distracted. Again, this is not to say chivalry has no place in these systems, as the core concept of connectivity should be bringing us together. Imagine how limited people would have been in terms of facilitating relations in older times compared to today. Back in those times, a person may never have got the chance to meet someone from another country during their lifetime, and today we have this ability at our fingertips. We have systems specific to dating where you scan through millions of profiles after entering personal criteria you believe yourself to be compatible with. Even with all of the advantages we gain as time goes on, to become more connected or globalized as a civilization, it seems as though we still struggle to be civilized. The relational behavior instigated over virtual connectivity can translate to our behavior in person. What is perpetuated as cool or relevant in this context would be harmless if it were to confined to the internet, but we tend to emulate what we see there in our real-world experience. More often than not, the celebrities and entertainers with major sway over these media do not live the life they portray, yet we follow suit in our everyday lives. This enables what has been recently dubbed the "bougie bitch" or "fuckboy" to reign supreme over their domain of your local nightclub VIP section. Although these environments certainly existed in their own respect long before the coming of connectivity media, it certainly seems to be intensifying the behavior involved with them and short-term relations.

Even with all of the information processing power we possess, we'll never be able to calculate a formula for long lasting relationships or chivalrous behavior. There is certainly something beyond all this connectivity to be learned from those simpler times in establishing an environment for lasting relations. The simplest teachings we acquire come from the experience uniquely tailored to each and every one of us—our personal feelings. One of the strongest emotions we have is love. The basis that love is a feeling falsifies all notions as to one's ability or competence, as everyone has the necessary apparatus in the mind, but rather points to one's capacity and how much one chooses to share love. Love is not easy, it takes strength to endure the trials and tribulations of love, and still there is no guarantee that you will ever find true love. Relations are not fulfilling because they are easy to maintain, but rather because the rewards come from hard work. As life goes on, you may find that your definition of love changes and what you thought was love was not, and what you thought was not turned out to be love. There is no objective to love, there is no end game, but rather the infatuation of the experience and the road traveled together. A capacity is referring to one's potential and how much one seizes it, and not necessarily being a need to pour wholehearted love to every person. This is proof that life's fulfillment comes from the value in true meaning and can be seen in creating something from nothing with purpose. Capacity is meaningless without purpose, and finding meaning is much easier said than done. Potential is most certainly a personal decision in basic preparation and seizure of opportunity. The same goes for matters of a subjective nature, and as it pertains to love and feelings, we display emotions. First, understand that feelings are not controlled, and we are subject to their will as much as you try to mask or divert them. Emotions, however, are how we choose to act out or emote those feelings, such as texting with "emojis."

Now, an example of converse levels of emotional capacity include being "deep" or "shallow." This is the classic example of depth vs. breadth—showing love and being loved deeply or widely. These media have inherently built in limiting and boosting factors of utility for both short- and long-term relations. Like swimming pools, we all have the capacity for both a shallow and a deep end, and this is proof of the importance of finding balance in life. Imagine swimming in this metaphorical pool; treading water in the deep end would be absolutely exhausting, but sitting in the shallow end would just be boring. In fact, optimized utility of what the pool has to offer comes from going back and forth between these ends. Those who tend to occupy more of their time on the deep side search for a comprehensive understanding and often entertain the perspective of others instead of their own. They tend to accentuate introverted traits and focus on the rationalization in the whole being greater than the sum of its parts. In general, people occupying the deep end tend to be more interested in long-term relationships. Those who tend to occupy more of their time on the shallow side search for a face value understanding and often entertain the perspective of their own impressions. They tend to accentuate extroverted traits and focus on what is to be attained by specific parts of the whole. In general, shallow people tend to be more interested in short-term relationships. Neither personality type is better than the other, and of course people have the capacity for free will to act in ways that are not expected of them in the ability for change. In general, you will find more affluence in throughout your relations by testing the waters of both sides every now and then to experience variety, or the spice of life. Regardless of where you perceive yourself to be the most comfortable in this metaphorical pool, it is not for anyone but yourself to judge one way or the other. This is especially important because you must accept yourself before you're able to understand yourself, and by extension, understanding and relating to others.

Media Energy Scale

In order to align your desired persona with that of the image you project to the world, you must first understand what shapes others' opinions about you. There is a particular model that stands out in combining all concepts of understanding and shaping public opinion. Aristotle's teachings of rhetorical appeal offer four aspects of being effectively persuasive or relatable to others.[1] This conceptual design is made possible with logos, pathos, ethos, and kairos—a representation referring to understanding, feeling, doing, and being. Understanding will coincide with logos, where you make a connection based on basic knowledge. This is mostly utilized as a bridging technique, as it will mainly serve as a mental function leading to further experience and empirical connection. Feeling will coincide with pathos, being that you will have formed a bond over something that you share sentimentally. This type of relativity is like a sixth sense in that it serves to illuminate matters beyond what is mental and is often referred to as intuition. Doing will coincide with ethos or one's credibility, a representation that you "walk the talk." This is very important because you need to establish trust, and a positive reputation in attaining a positive reliance. Being, will coincide to kairos or timing. Timing is everything; being at the right place at the right time can alter the entire course of your life. Being, in terms of the rhetorical approach to controlling your vibrations, is not only a literal state but figurative as well. Being there for someone is most commonly associated with an obligation to their well-being or simply being in touch however you can, especially in a time when being does not have to be in person. A full rhetorical intuition is the second nature that serves as the power of love, allowing us to navigate energy and identify potential outcomes. As the expression "on the same page" suggests, you find yourself completely in sync with another person on all levels of Aristotle's rhetorical model. There are many signals that our body and mind use to communicate with one another, some being more apparent than others. Having a physical presence over these media is a different form of communication because it is selective to their limits as a platform. This may have numerous side effects on how we selectively choose to present ourselves and in the ways of interactions with others. It is important to remain centered among these media and not allow a perceived limitation of these devices to translate into a limit of yourself or your opinion of others. We need to remember that people are far more complex than these systems are able to capture, so keep an open mind as to what another person is actually experiencing beyond their virtual persona. Overemphasized importance in media representation among relations can be a true coup de grace of chivalry, but it is not yet dead.

Love is a universal language in relations; however, media outlets require a specific colloquialism to pontificate clearly through to achieve relational success. Multiple translations applicable to the idea of communication exist such as the classic visual, auditory, and kinesthetic (VAK) model.[2] With regard to natural real-world communication, we see a comprehensive form in body language, one of the more commonly noted ways used to decipher these classic messages in relation, often missing in media. The major condition of this language, and many others among the classic model, is that we do not always choose what these messages are as they are more instinctive. The subtle nuances of in-person natural communication are not easily replicated over virtual channels. Due to the subjective nature of certain signals being immeasurable, we must interpret them subjectively through our intuition. On the contrary, we see measurable records in the ways we project to the countless channels of media limited to a fairly objective interpretation. Even if the message can be read and interpreted clear as crystal, there is exclusion of feeling that is lost in translation and can be better understood in observing the natural world. Our body and mind subconsciously project signals in our reactions that unintentionally communicate to others in our immediate surroundings. In particular, the eyes are one of the most prevalent of the conventional five senses with regards to the subconscious, and the eyes never lie. There is a saying that the eyes are the windows to the soul, suggesting that visual stimulus is more than just eye candy and perhaps entails a deeper connection. In this respect, media use can be compared to wearing sunglasses indoors, where in certain contexts, it serves no other purpose than to shield insecurity within social interaction. There is truth to this in respect to how different a person acts when natural stimuli in communication, like eye contact, is lost. This, among many other stimuli, is why conversation tends to be more comprehensive in a face-to-face environment as opposed to over the internet or phone. When speaking to one another, there is so much to the words in natural communication—timing, intonation, and so on—that is lost when replicated over plain text. It only makes sense that in a time when the majority of communication stimulus is inanimate, we feel the need to overcompensate for this loss. Over this millennium through to today and in the future, people will have communicated without face-to-face interaction more than ever. Mail turned into email, calls turned into texts, and communication became more and more simplified, thus losing the subconscious connection of sensory stimuli. This has begun to make a comeback with increasing technologic capabilities such as video calling, but even then, the interpersonal effect of face-to-face palpable energy is lost. There's no question as to how certain physical contact actually establishes trust. When meeting someone, a handshake is all it takes to establish a sense of common ground. Take the idea of video conferences in business as an alternative to actual office meetings and the effects of missing elements in basic interaction. In relationship terms, this type of communication is imperative, and unlike a business handshake, the interaction purpose has no objective incentive and depends upon sharing pure feelings to succeed. When two people in love look into each other's eyes and relax, their heartbeats often synchronize. Ultimately, it is in our nature to connect, just not exclusively in the way of our current involvement with media. The capabilities within modern devices do not perfectly recreate the more subtle nuances of natural communication, but they do provide other utility for better relations. The point is not the technology being debilitating, but rather in how we use it. We need to understand that new technology is not a replacement for natural human interaction; it should be used to enhance it.

Although we see different levels in variance and similarities between natural and technical communication, they are perfectly capable of coexisting. These two forms of communication should be balanced in their utility, even promoting one another, but they should not be given the same value. Apropos, an "I love you" text can be just as powerful as saying the words in person given the right circumstances; however, an "ily" text and an "i love you" text are not the same. Science has shown that there are neural connections like a web in our brains, and certain thoughts trigger these billions of neurons emitting electricity to fire and send messages in the form of waves.[3] This can literally be measured as electromagnetic brain waves, whereby simply thinking certain thoughts produces consistent wave frequencies. These waves are measured in order from dormant to active in the form of hertz (Hz), the unit measuring frequency. In terms of activity, the frequencies are respectively ordered from high to low: Gamma, Beta, Alpha, Theta, and Delta waves. This is an example of what empirical studies have proven to be present in natural communication, but there is yet to be any type of energy discovered in communication over the channels within the networks of cyberspace. This being said, there is a ton of energy still flowing through these channels, and although they have not been exactly classified yet, they do hold significant value for the purposes of communication. Regardless of the type of signal, whether conscious or unconscious, measurable or immeasurable, it is undeniable that energy is there and interpretable in any social setting. The scale of perception can be represented on a graph of x and y axes that are either subjective or objective. A natural environment is more conductive to subjective communication, as it is based on the mood of the moment. When we are confined to a technological environment, we tend to use more objective communication as our messages are concrete and direct. The part in conjunction with having healthy communication is more reliant on the third dimension, or the z axis, measuring extent of energy exertion. For the purpose of communication, picture this axis to measure the extent with optimism being on the positive numbers and pessimism being on the negative numbers, where zero would be indifference. The reason this aspect is the most important is because it is impossible to choose the input that affects the subjective or objective stimulus that you receive in life; however, thanks to reason and personal discretion, it is your choice as to how you react to this stimulus, either in a positive or negative way. The nature of our DNA is inherently different from any other person, so we are born with various traits that can prove to be advantageous or detrimental to our relations. The more notable part relating to this z axis is how we also have a unique perception of stimulus throughout our lives, known as nurture. The point here is that you have the capability to take nature and nurture and choose what you make of it. This is due to a complex wiring of the human mind, enabling the unique feature of critical thinking most commonly referred to as "free will." These media systems may prove to be influential, but they are obsolete without the user end in control, so we are able to make the ultimate decision as to what role of importance they play in our lives. A healthy life and relation will always contain balance, so it is best not to lean too heavily either way and see utility in both natural and technical communication.

Chivalrous Morality

Observation of free will through media interface highlights the process in which we choose to create a separate yet congruent reality of our true persona. It is important to make the distinction that these media interfaces are isolated, although relevant to reality because they are compatible to both forms of human perception. Although these systems are a simplified mold or framework for who we truly are, they still bear significance to our unique identity. This theory is dependent on the fact that there is one true reality that depends on an infinite amount of dynamics surrounding any given matter, which it is impossible for the mind to comprehend. This is why there are always three sides to a story—what you think, what they think, and the truth. The world as we know it is such that there are actually much more than 50 shades of gray, but an infinite amount of color, and without the existence of abstract concepts, there would just be objectivity, or black and white. Having said this, it is important to be able to distinguish the aspects of your natural being and the virtual representations over these systems, or 0s and 1s. If you overemphasize the importance of these online profiles, you will lose touch with validation in the subjective nature of your being, and your relationships will not have balance. In objectivity, there is a basic assumption that we are wired to compete, to gratify our own desires, and be self-interest machines. In this regard, being biologically fit is confined to the ability to find a mate with comparable strength in accordance with Darwin's theory of natural selection. As our society points toward a perpetual state of competition, this is not actually natural, and even Darwin's theories are oftentimes misinterpreted. Given the choice of interaction between humankind being either to dominate or cooperate, the one that displays the closest thing to love is always triumphant in nature. In regards to connectivity media, our societal standing justifies competition in having the most likes, comments, and shares because they promote independence and self-interest. However, in nature, this is not effective; you actually see the subjective concept of orchestrating coordination for the greater good. Kindness and selflessness will get you further in relations than any imposition of control. When Darwin wrote the *Descent of Man*, he mentioned the survival of the fittest twice, but the word *love* 95 times.[4] When Darwin wrote his first book about human nature, he stated that we're not very fast, we don't have big fangs, we don't have the muscle mass of gorillas, but what we do have is the ability to cooperate and to take care of others, and that sympathetic advantage is the strongest feature of our nature. So far, we have observed that the more advanced cognitive animals share a common brain function called the mirrored neuron. This means that when observing an action performed by similar species, the same neuron fires when performing the action itself. This lack of ability to distinguish a self is why we feel love when we see love, we feel pain when we see pain, and is the basic fundamental apparatus in functional compassion. The point is that these systems are not good or bad, but what we make of them, and what it really comes down to is your will to be open-minded. This does not necessarily have the connotation of being an optimist or a pessimist, but rather the level of flexibility you allow your mind to practice. This ability is extremely important as it allows you to have the perception of not being concerned with the glass being half empty or half full, but more so with what's inside the glass, what the glass is made of, and what purpose the glass serves. Likewise, these systems are based on a framework of objectivity, but media can be used to inspire subjectivity and fuel our relations.

In subjective reflection, these media systems provide us with the tools to build upon relational character identity through manifesting the law of attraction. Your self-identification is the basis for all relations, as knowing yourself is a prerequisite to loving yourself—both of which are necessary for engaging in and retaining love with another. Many things concerned with connectivity account for "peer pressure," but if you know yourself, you will maintain your standards instead of letting external forces change you. You must allow a popular trend or glorified lifestyle to exist in peace in its own respect and mind your own matters to prevent these from taking hold of what makes you. In this day and age, it is inevitable to come across such pressures that seek to change you, and it can be easy to succumb as to temporarily ease any stress it may cause. However, as beneficial as it may prove in the short-term, the long-term solution is apparent in learning to stay true to yourself and escape with dignity. In regards to relations, this knowledge will serve to combat the practice of lowering your standards or finding insecurities. If you know yourself and are happy with the way you are living, there will be little room for interference. On the contrary, if you are unsure of yourself, you will manifest toxicity that will only voice your perceived weakness and lead you to unhealthy relations. Self-control and stable demeanor are the instruments of creating your desired reality and clear conscience; consistent emotion and rational perception will help emit the energy that you want to see in others. People will subconsciously pick up your energy, and you will vicariously control the way others affect you. To that point, you must understand that there will always be things beyond your control and things that undeniably control portions of your life. These media systems offer a very low degree of control as they are a simulation of reality preset to specific criteria. The sooner you accept this, the easier you will be able to facilitate these things into forming something productive. It's good to be conscious of how others feel about you, but there is a point where it is unhealthy and you care to a fault.

When you crave acceptance to that extreme, you change yourself, and that is where you need to draw the line. The main reason attesting to the importance of this principle is the concept of projection, being that what you see most clearly is really a reflection of who you are. More often than not, the things you love or hate in other people or the world in general are the very same things you love or hate about yourself. This projection of truth will create standards for others to base their perception of your relational significance, and we can utilize these outlets as a channel of energy for the law of attraction. In relation to loving yourself, you are the only person you should feel the need to impress. If you do not love something about yourself, you are the only person who can change that. This is contingent upon the type of energy you emit into the universe, which is returned to you through the law of attraction. The law of attraction is similar to the concept of cause and effect in which nothing happens without a catalyst and every action will produce an equal and opposite reaction. Upon quantitative systems in this media, these actions and reactions are more easily measured, so a level of care is imperative. There is no such thing as coincidence; everything happens for a reason, and relations are derived from choices you have made in your projected image as a result of these actions. Therefore, the law of attraction is not something that you should interpret in any sense of objectivity, because it is something that requires the perception of abstract energy.

Chapter Two

The Millennial World

Sociological Relativity

Technological media is meant to be used as an experience enhancement, not a replacement or means of disconnect from our immediate surroundings. It's no secret that the key to a successful relationship is communication, but this is complicated by the fact that the field of communication has changed so drastically in the past 10 years. As an early '90s baby, I was raised with the invention of many foundational building blocks in technological communication. We were the ones who made the transition from passing notes to texting in class. These breakthroughs happened so fast that children quickly became more adept in their function than their parents. This accelerated growth rate of connectivity left a vast majority of people without the proper guidance in appropriate application. Almost none of these capabilities were in existence with the generation before ours, and our elders seem to be in a state of hopelessness in search for any type of common ground. Not only are the attitudes toward these capabilities drastically disproportionate between a single generation, but compare the societal ideas of today to those of the golden age of chivalry. Just one of these lost concepts includes the idea of participation, or actually being a part of the immediate world in front of you. There are definite times when the use of these media is counterproductive to your relations, usually including a focus of social events that have precedence without such systems. Countless examples from social situations display the disinterest we show in each other's presence, where hanging out with friends turns into sitting around with our eyes glued to a screen—physically present but mentally far, far away. We see this everywhere— media bombarding us from the streets of Times Square to the junk box in our email account. Not only has the public eye adopted this habituation of media inclusion, but private settings as well, such as dining restaurants putting up flat screens all over their walls. This blind inheritance of multimedia stimulation often serves to enable a distraction in relation only to disconnect from each other. The power of engagement with the real world is beginning to go unseen, so it is imperative we fight the urge to sit back and watch it all play out, and rather participate in the constitution of relational facilitating environment. Establishing rules for yourself and others such as not answering calls and texts at a dinner table can help in separating time for both types of engagement in their appropriate setting. Like most things, these advances are harmless when used appropriately, but there are some that tiptoe much closer to the line in promoting behavior counterproductive to these causes, including chivalry. The first problem with society's views of chivalry is that it is almost become completely out of sight. What's worse is that it's not just lost—chivalry has been gone for so long that few people even know what it looks like anymore.

Media influence has facilitated a new era of how we define the characteristics of ladies and gentlemen, and by extension the very principles of chivalry. Today, we see gentlemen, and furthermore chivalry, in a very different light than we perceived it at its peak in the Middle Ages, which calls for clarity on what it truly is. This being said, the effects are not exclusive to a man's cause, as women are often the targets and provocateurs of this change. TV series such as *Teen Mom* or *16 and Pregnant* glorifying such behavior are all but helpful to the cause of chivalry. Chivalry is a concept that should be transcendent to genders or measures of time, so it begs the question how it got to this point. How is it that we refer to a strip club, a place of objectifying women, as a "gentlemen's club?" What we have been conditioned with in current society to understand and accept as normal is in fact an oxymoron such as that. With regards to the media's effect on relations, it is nothing new to see that sex sells, but sex, love, and the relational process have been drastically manipulated to fit their agenda. We can see a major media endorsement of this condition and the idea of reality TV, shown in a popular program *The Bachelor* and its counterpart *The Bachelorette*. The very idea of these shows is based on someone taking their pick from a group of contestants to share a relationship. This show is a mockery of the relational process as it poses the part of being real, but it also sends a poor message in the advocacy of one person literally playing at others like a game to see who wins. This is obviously not how we naturally act, but for some reason when we see it on television or other media it then becomes acceptable. The fact that this agenda is so successful in creating a playboy/playgirl speaks volumes to where our generation's priorities lie. There is an underlying message here in promoting an unattainable eccentric lifestyle, where in reality, any reasonable relationship requires only the bare essentials in sharing feelings and not all the hype that comes with broadcasted drama. In contrast to these manufactured icons and celebrities, a true lady or gentleman has no need for this exposure in display of humility, and the picture-perfect knight in shining armor is mostly confined to dreamy fairy tales. A testament to internal validation is how the crest does not make the knight; the knight makes the crest. In fact, the more capricious and impulsive persona comes with the heightened risk of lacking the capacity for sentiment. Suffice it to say that boorish characteristics are not particularly amicable with chivalry, and we must find prudence to combat such ways. A major contribution to the ways of righteousness and humility in knights is due to their roots of affiliated Christian religion. Among the standard Ten Commandments against sin, we see actions that produce positivity and spread love. In the ideals that coincide with this religion regarding hubris and indulgence, it is only appropriate that this media endorsed behavior is seen in coinciding with chivalry in Jesus's words, *"It is easier for a camel to go through the eye of a needle than for a rich person to enter the kingdom of God"* (Matt. 19:24). Alter the concept of selflessness here to a relational respect and you will find yourself a knight in the kingdom of chivalry. This is not to say that being a wealthy person is wrong, but the temptations brought by vanity and overindulgence lead only to an imprisonment of body and mind. This technology can be a limiting factor to natural engagement, but can also prove beneficial to overall relational success when used correctly. We just need to concentrate on the power of natural engagement when media use is not appropriate.

How we value ourselves and another is a reflection of our ability to coincide and coexist with each other and is not to be confused with sheer popularity. Popularity is a one-way street and does not serve relations as well as two-way communication, cohabitation, cooperation, or the many other co- root words. The millennial world has undeniable obsessions with our newfound connectivity and who or what is popular, as trending has literally taken on a new meaning quantified by the number of hashtags associated with the topic. In conjecture with our obsession to keep up with the Kardashians or be relevant, a recently coined term, "meme," proves to express our entertainment by the ridiculous that comes as result of the media. Some of these memes include mere jokes that are concerned with pop culture, but others portray general attentiveness to more serious matters. Examples include pictures of our predominant hand literally being an iPhone or the tragedy that trees only produce oxygen instead of Wi-Fi signals. All jokes aside, some of these things are not that cool as there is truth in jest, and these things seem to be getting funnier or more true. Humankind has always been subject to thralldom by material, but it would seem that we are shifting toward subjugation under ideas prompting a lifestyle that projects an image of false reality, popularized in "fake it until you make it" and "fake news." This has been historically dangerous as Adolf Hitler used *lügenpresse* to promote a Nazi agenda. People are now actually embracing the idea of a false idea until it gains enough support to become viable. The obvious problem with this is that the perception is a mirage, and when put to the test of functionality that an authentic version normally withstands, it will falter. A perfect personification of this concept is seen in reality TV. This is especially concerning as the vice of an idea comes with chains more difficult to be liberated from than those of a physical being. The word for this type of oppression is *fame*, and the spotlight has become increasingly prevalent beyond the general media into the narrowed spectrum of social media. This type of connectivity is alluring as it comes with the ease of being loved widely, but is deficient in the capacity of being loved deeply. Not only does it lack an important distinction in deep love, but fame is kin to infamy and may lead to negativity. Proof of such can be seen in the lack of quality or wholeness, only to be replaced by quantity and a superficiality. A catchphrase, the validity of this search in modern times, is in having "15 minutes of fame" as opposed to a fulfilled life of humility. These are the disposable attributes of society that devalue people in being temporary to our lives until something else comes along to steal the show. The fling that makes the cover of major label magazines only promotes their profitable agenda as if these relationships were just publicity stunts, because after all, any press is good press, right? There are a handful of people chosen to represent certain niche markets that have become so famous that they often lose touch with reality. It does not stop at well-known celebrities, but begins to enter a microlevel in small community application to exile all walks of life not considered to be similar to those deemed famous. Never in history has there been such glorification of public figures that are actually famous for the worst of reasons in regards to adhering to moral and ethical codes. Not only are the reasons that we love these people so depraved, but the love itself is hollow and promotes a disillusion of the word. Fortunately, it is not necessarily the fault of these individuals as they are just products of their environments, like test subjects in a new invention with no precedent or guidance. In order to repeal this behavior, like a parent teaching a child proper manners, there should be guidance to etiquette from those who have more experience, and this is where millennials hold a place in having inherited responsibility.

Netiquette and Techlination

Much like how we behave in person, there is a certain set of manners and acceptable conduct to abide by when deliberating our actions over these media. This is the golden rule: If you wouldn't do something in person, don't do it over media. However, there are times when in-person actions are tech-inclined, such as browsing media at the dinner table. In this sense of the word, media encompasses both the physical and virtual implications that come with the many forms of their utilization. When mingling these capabilities with real-world interaction, there are certain ways to avoid unintentionally offending others, which mostly comes down to timing. There is a time and place for everything, and obviously mixing actions such as texting and driving is considered dangerous and inappropriate. It does not have to be this drastic but bad habits like constantly being in and out of attention to your immediate environment can be detrimental to your relations. This being said, in relations, the actions do not have to pose immediate and direct effect while still being bad etiquette. This can also be advantageous depending on how you use it to your advantage. There is a concept of psychology defined as code switching; in this, context would be described as behavioral adjustments in adapting to your environment through use of media. People tend to speak and act differently in different contexts, and as such, we change our behavior in the same fashion over the internet and other forms of media. In the same way that you do not speak to your parents like you speak to your friends, we do not act face-to-face as we act over the internet. This is to say that you probably would not print out your Twitter feed to present as a resume for a job application, and that is how we must perceive these actions. They are not a definitive reflection of our whole being, but rather an expression of ourselves that coincides with the applicable environment. You have to understand that the media and our communication systems have a very different vibe, in a sort of anything goes structure. However separate of an environment these media may be, the actions and expressions we portray translate to our real lives. Some things come simply by nature, and we should leave them this way without imposing our technological capabilities. It is useful to ask yourself how the proposed action will affect your relations as a whole, and whether it truly reflects the nature of your character as a human instead of what it may do for your virtual representation or profile. With regards to netiquette, it is obvious that the invention of the internet has brought about a separate global culture which bears the obstacle of being sensitive to respective geographic cultures and the laws of their lands. Something normal in one place may be unacceptable in another, and this is relative to basic circumstantial context—however, the internet does not offer such clear-cut cues as to proper etiquette among local environments. This brings to question if there should be some form of statute for all users in some kind of separate global government, or if citizens should rely on their local government to set these boundaries, if at all. Laws are certainly objective, but when dealing with something subjective such as relations, this matter gets even more complex. This is why we must think in terms of civility and basic human rights in unified global culture for use of discretion in avoiding the separation of ethics between our actions on the internet and real life.

The purpose of this media is not to facilitate speculative judgment upon each other, but rather to embrace and illuminate our differences and similarities. When certain people generate attention, it is a common practice to either criticize or canonize them and their apparent ideologies or lifestyle. It is not to say that these media systems are responsible, but more so how it sheds light on things and as an effect, what we are enabled to see about one another. An important aspect of this concept is where we are able to make a distinction between celebrity and philanthropy. Celebrity is a very common type of notoriety in millennial society, in basically achieving attention that focuses on the things you have done for yourself. This is the framework for the idea of fame, whereas a philanthropist receives notoriety for selflessness and noble deeds. It seems that today's global society on media has shifted to inheriting this celebrity type of endorsement because in this day and age, it has become very easy to generate a considerable amount of fame. This is mostly because these media provoke an obsession with ego and self-centered promotion. In earlier times, notoriety for philanthropy was in higher regard, but unlike celebrity, philanthropy will never be easily attained. This social status is awarded to people who earn it achieving breakthroughs for the greater good, and not given to people who sell their soul for 15 minutes of fame, or worse a lifetime. It is a wonder in itself how any of our peers will be able to become a person of reputable warrant with the amount of blackmail or smear campaign potential circulating in plain sight. This is major concern, but in reality, a more serious concern should be focused on the fact that what society once thought worthy of blackmail will become the norm, or even be seen as noble or admirable and rewarded with such a high regard. This trend is nothing new as democracy is essentially a form of popularity contest, but where do we draw the line of decency among the relations we hold between our personal lives and relations. Of course, times are and will always be changing, but certain concepts such as seeking dignified public figures, and the concepts surrounding how we see one another in dignity should remain constant through time. This type of famous love is merely an adult version of schoolyard infatuation, where you adore someone for the image that they portray, instead of the sentiment in their actions. Popularity is a superficial form of care and should not be mistakenly substituted for love, or for the purpose of this book, share the quality of chivalry. The sheer amount of accessibility, along with the lack of effort involved in having influence over the masses among these media is what dilutes encouragement of individuality. This leads to a common mindset of heightened importance in immaterial matters of the present, like how we prefer to take pictures of our meal on a date instead of enjoying the immediate company, all to appease people who are not present and most likely do not care as much as the people we are with. Much too often we go out of our way to seek the approval of others to the extent that it is actually counterproductive to them and ourselves. Modern connectivity comes with abilities that allow you to be a part of someone's life with a touch of a button, but this is just low hanging fruit. This concept that nothing good comes easy reinforces why a solid, true relationship rarely comes from an app that requires a selfie, a bio sentence, and the ability to swipe left or right.

If used in excess, these media can be enabling to our addictions of ego and self-promotion, in place of moderation and honest self-realization in one another.

When you learn to love yourself, you break barriers of your insecurities and are free to recognize your being as more involved than separated to your fellow humankind. Loving yourself will prove to be easier said than done, as sometimes the things that make you unique can be debilitating—but, *c'est la vie*. You will also find that some of the things that were once debilitating, placed in a different circumstance, can be empowering—and that's life too. The key is your ability to adapt, and it is within the things you bear in signatures that make you unique. This media has presented us with an enormous potential for all things to be expressed and created, which can apply to relations as well. Energy goes where energy flows. It is important to emit positive vibes when you are in connection with your friends and loved ones, and these systems provide just the right kick to boost your relations in this regard. This being said, it can also be dangerous in the extreme when you depend on the validation of others. We tend to carry this out by measuring love like a statistic, but this will only complicate your perception of your relational success. In this case, you begin to hold your own expectations against the things others have and inadvertently limit yourself. This ambiguity of love and hate is called jealousy, and it can lead you to attempt a change of who you are modeled after someone else. When loving yourself, you enable the ability to be happy for others and share in their own unique success. A condition being attention deficit disorder (ADD) is a perfect display of how we have been stimulated with this kind of media recognition and emphasis on objective measures and insecurity. This was just recognized within the 1900s, only to become widespread throughout our generation—and this is by no means a coincidence. It was a result of our unprecedented input of stimuli from the advancements in technological media focus and entertainment. Our fellow classmates were prescribed a synthetic comparable to cocaine in order to concentrate on schoolwork. We thought there was something terribly wrong with our brains, when in reality, it was because we were comparing test scores and quantifiable data instead of understanding the issue.[1] We were raised in an era of youth who lacked patience, while mastering the art of procrastination. This social behavior is a major separation between the attitudes and lifestyles that can be noted with previous generations. However, it is not always so obvious to those individuals caught in the movement as we are somewhat products of our environment. These media can be seen as a form of drug with its addictive qualities, but it's not necessarily a dangerous thing as long as we practice moderation. Understanding when excess has been reached enables you to act against anything harmful to you or another's health and well-being. The strongest form of controlled substance is the addiction to an idea, something intangible that cannot be taken away from you like a scar on the mind. This is why it is important to remain open minded among these media, and to not give in to the ego trip as a product of their objective utility.

Objectified mentalities can be perpetuated by media promotion of quantified importance, but it can also shed light on subjective beauty of human nature. Today's generation of young adults, or millennials, have used these technological capabilities as a prolific source and outlet for creativity. There are almost no limitations as to the ways we can express our interests over these media systems. Not only are we able to create virtually anything, but the beautiful part is in how we are able to share with one another. This is the source of power for creative development in our means to build off of what others have done before us; there is no invention without innovation. With respect to relations, this allows for effective and efficient facilitation of creative tactics in fueling what brings us together. Being able to share is the most powerful tool available to relations, and now we have the ability to share life in more ways than ever. We see this in the definitive term *viral* and how instantaneously a single act caught on camera can bring light to someone's day, and a bond is strengthened through shared laughter. There are so many good things to come out of the simplest of gestures, and we can use this media to act as an agent in our relations. This being said, we still need to act responsibly and focus on positivity because media bears a fine line between productive and destructive. However creatively stimulating as these may seem on the surface, they have equal potential in posing a danger to those who stumble upon something they weren't looking for. Today's platforms of connectivity offer an entire world of information and media at your fingertips, and in certain respects can be notably comparable to the Wild West out there. Even with the progression of surveillance and internet security, there is still a "dark" or "deep" web that is under the radar and out of conventional reach. The internet almost definitely has a place where you can see anything, buy anything, even do anything to the extent of virtual reality. The internet truly has no chill, and there are hardly any rules or regulations like there are with other forms of media such as the FCC and the SEC government agencies. Kids from around the world are literally hacking into Fortune 500 companies' homepages just for fun. This is not to say these systems advocate this extreme and illegal behavior, but they do offer the opportunity for these acts. With respect to our social media, this potential of unlimited reach to one another can serve to make or break our relations. A lack of censorship can lead to a desensitization of reality, not necessarily to the point of life and death, but in general it certainly affects our relations. Of course, relations are a thing of subjectivity, so your actions have potential to attract and repel different crowds. This is why it is important to remain inclusive in what you convey as to convey positivity for both those who are interested and willing to engage, or a respectful pass for those who are not. Relations depend on a source of inspiration, so we should utilize these media in a way that is innovative as fuel for the impetus in our subjective passion and expression.

Millennial Timeline

Speculation of the event horizon with these media on the millennial timeline shows the full effect of their impact on our societal and relational dispositions. In general, I am speaking for my generation of millennials, and exactly how these events were timed perfectly to coincide with our early childhood and adolescent upbringing. We were the ones with the very first taste of social media in all of its forms and were dropped into it with absolutely no supervision, resulting with a degree of expedited maturity. This being said, we were also part of a time when many of these things did not exist, and we can clearly see the technological differences between our childhood and children of the future. The first full internet service on cell phones was introduced in 1999, and following close after came various outlets of technological media.[2] At this point, we were barely teens, but until then we relied on more classic forms of entertainment, whereas today's children half of our age at that time have full access to the devices and media we use as adults today. Today, there are kindergarteners who can barely spell but can easily rock you on the latest video game. Our brains were in peak development stages as we matured alongside its evolution through different stages of increased connectivity. It would seem that the timing of my peers' induction to this new world was truly a one of a kind chance phenomenon—just the right place, just the right time, and just the right socio-environment served as an oasis of a test tube to draw a reaction. Immeasurable power was in the palms of our hands and we had no idea as we steadily dabbled in programming language, coding just to change the aesthetics of our profiles. These newly acquired abilities were used in creating young sensations, overnight celebrity, viral videos, and the potential for much more. A timeline will provide contextual perspective to when connectivity first started changing and new facets to relations evolved into what we are accustomed to today. Take for instance the invention of the telephone, soon thereafter entering the realm of relationships when phone sex was picked up. The same concepts were adopted when cell phones came along in sexting; now think of the capabilities brought by Snapchat and FaceTime. Regardless of your personal preferences, these interactions should be distinguished for what they are, because they are certainly not replacements for reality. Social networks are just networks, which are inhibitors that synthesize and facilitate, but they do not take the place of natural interaction. The types of natural interaction outside the confines of technological connectivity are boundlessly different from these platforms. The technological forms of connectivity and social networking lack the subtle nuances of organic communication and can often lead to overlapping perceptions. If you're having a conversation face-to-face and someone doesn't respond for two days, it is rude; whereas online, this is often normal. This demonstrates how mixing ideas of the two environments can lead to confusion of what is acceptable relational behavior. Being born without these capabilities only to inherit them at an appropriate time proved to be advantageous in how we value technological connectivity proportionally to old-school forms of communication and interaction. Our generation put us in the perfect position to have perception of the entirety of these media and the effect of what they currently are.

From the very beginning of our childhood development, we were matched with systems that not even our parents knew how to operate. In 2005, I entered middle school in the sixth grade, and at the age of 13, I already had a cell phone and a personal laptop. As fate would have it, 2005 was also the year that the recently founded website Myspace became the most visited website in the world at several points in its heyday.[3] In entering the sixth grade, we were amidst a new world of puberty-driven behavior with the revolutionary addition of having a virtual playground to host these activities. The inclusivity of outlets offered unmediated platforms to connect with anyone and everyone who was also partaking in the connectivity of these media, and this was only the beginning. Myspace will likely be remembered as the first social media with its user-friendly interface and profile customization features such as the "top 8" where you chose which friends to put on display as your best friends. This feature was a huge advancement for how we saw our relations because you could objectively measure who you preferred to associate with. This objective approach to social information sharing was welcomed with open arms. There was very little held back, and all too often in the ways of indecency, as popular kids would reign supreme, and the less fortunate ones were torn to shreds by internet bullies. Perhaps even worse, unpopular kids were simply disregarded as unimportant or uninvited with a "you can't sit with us" type of disdain. Regardless of the passive-aggressive or blatantly direct tactics, there was nothing like it in the form of free speech—and mind you, we were still learning to say please and thank you. Rudimentary systems of this time provided outlets of expression that were utilized in a foolhardy manner, but we were only products of our environment. It was the dawn of these forms of connectivity, and we had no precedent to serve as a structure for what was the norm. It was through no fault of our own that we took to the wind in such a manner, which shows the importance of transferring our experience to future generations. Needless to say, these systems were powerful by nature, but being bred by them created a new class of thinking that our parents did not experience. Not only did they not have these capabilities at their age, but they could not take part at the time because they had no place among us. This allowed the youth to hold a majority influence over the behavioral norms associated with these media systems. We held control over these media, it was our time, and we continued to adapt and mold them as they continued to adapt and mold us.

As time went on, the social media concept gained more traction, and one of our peers quickly became a billionaire through his contribution—Facebook. At the same time that I began my freshman year of high school, media paralleled our culmination with a higher degree of connectivity in the social network. In 2008, Facebook had taken the throne as being the most visited social media website in the world.[4] One of the site's defining characteristics is that it started out as an exclusive or private network, essentially a blog strictly for a select group of Ivy League university students. The aim was directed toward a localized level of sharing and private social event organization. The day after a party, you could check back to the event page and have access to any details you may have missed out on in the form of posted pictures, and whatever gossip one could possibly ask for. Shortly thereafter, Facebook expanded parameters allowing anyone to make a profile, enabling open media gossip and an inclusive social dynamic. As time went on, it took on a new format for uploading any kind of content including links to other websites accompanied by a live activity feed between your connections. The ease of accessing the latest news on current affairs was at a peak, but it wasn't exactly aimed at information. At the time, this was not what it is today, and the dynamic of a multibillion dollar enterprise was not the main focus. As this began primarily as a social media, the majority of what we were sharing was concerned with regards to our friends and social associates. Back in its conception, it was distinguishable as a tool for personal relations, whereas today it is flooded with random people and sponsors among corporate agenda through all things consumer-related. The wealth of information beyond anything directly applicable to our lives or relations is astonishing. There are currently features in minute interface options and niche forms of accessibility that extend beyond utility of socialization. Today, this platform is not just a social media, but a system of networking, data collection, and entertainment that is beyond even the label of multimedia. It used to be a simpler and seamless environment with its simplistic features and general aide in the ways of relational functionality. Perhaps its crowning feature at conception was the relationship information offered, as it reserved a space in your profile's general information for being "Single," "Complicated," or "In a Relationship." Not only did this serve to clarify the status of any given person's relation, but it provided a hyperlink to whoever you were dating. This was a major step for social networking because you could follow the details of personal relationships through mutual friends and how people were connecting.

Social media continued to keep up with tech advancements, meeting the demand for networking adaptations with a platform for picture sharing. In 2010, another big player in social networking came along, but this time it originated as a phone application. This new type of media offered a profile composed purely of pictures—each separately accompanied by a caption, a geotag, and a usertag. Instagram became successful for various reasons, one of which being the simplicity of the format. This was a similar factor in why Facebook was so successful, so people obviously responded well, especially after Facebook bought it for $1 billion.[5] In this simplicity were the attributes of the profile, being that you had a header including nothing but your name, picture, bio, and most importantly a rather conspicuous indicator of how many people you followed and how many people followed you. The most notable effect this introduced to social media was in the inherent competition and general recognition to be gained from having the most followers or at least a decent ratio. This created somewhat of a blurry line between the importance of seeing people for who they associated with and how many people they associated with. This is yet another testament to the effects of media objectifying our relations, and we literally bought into it, purchasing followers to look more popular. This led to objective based changes in how people would do things they don't really like, just to get likes from people they don't really know, as if to advance their fictitious celebrity career. This is not to say this was all bad, as many people with actual talents gained deserved recognition, but there are also those who sought to sell out and attain a shock value following for absurd or lewd behavior. It is not always as clear-cut as to what is socially acceptable, due to a figurative and literal application of filters. This was perfect for taking things out of context and showing the more appealing side of what may have been an ugly situation. We later received an evolution of photo sharing in the application Snapchat, providing a platform to broadcast every thinkable aspect of personal life in the most casual demeanor. The difference with this application is that the content is temporary, and theoretically there would be no trace of what you did after 24 hours. This affected relations beyond the prior attitudes toward too much information (TMI) as we started sharing TMI over media, and discretion went out of the window. People's reputations are sullied by themselves, and in some extreme cases, these people are followed and made famous for the worst reasons, often embracing this fame as if to be bragging about their crudeness. This is just an example among the many ways certain fame can turn to infamy, all for their 15 minutes. This short lifespan concept in social media reinforces the idea that our society advocates for temporary and disposable aspects of our relational behavior. This translates to a hesitant mentality toward commitment, which can be noted by the new generational hookup culture.

Dating/Hookup Culture
Social media began to directly affect how we value and

conduct our relationships when they were made for the sole purpose of facilitating them. In its nature to evolve alongside our generation, social media gave way to another phone application in my first year of college in 2012, only this time it zeroed in on fixing you up with a date. This was one of the first initiatives toward being primarily relationship-based as opposed to the conventional casual environment. This change was made possible with the invention of the phone app Tinder, the media platform officially enabling my generation to adopt a hookup culture. Think of what this meant for me and my fellow classmates. I was a freshman among 100,000 other college kids on campus or within a few square miles of each other attending a legendary party school, only now we had an app that showed everyone around us was looking for the same thing. Based on your parameter settings, this app compiled a list in the thousands of potential matches. The way you were presented with and to potential matches was by a profile picture and a name, which you then swipe left for no or right for yes. Obviously, most people were just using this app as a means for finding a one-night stand instead of a long-term relationship, but therein lies the problem. This was a perpetuation of the expendable disposition toward relations and reducing a human being down to a name and face that could be passed along to your liking. We had inherited an acceptance to treat one another as temporary and disposable, only to realize soon thereafter that we ourselves had become the victim. This app concept in anonymous introduction continues progression with the likes of missed connections, or recalling a moment when you entered GPS proximity but missed the chance to introduce yourself to another user. This means that being in the right place at the right time is now grounds to initiate a relationship. This encompasses the main principle of what is wrong with these apps, in that it just promotes laziness. This is not to say finding a relationship should be a challenge, but there is not a whole lot of room for chivalry in such basic practice and dependence on an app to set you up. Not only do these apps shortcut chivalry, but their promiscuous nature can put you in a bit of a bad predicament. A conflict can arise simply from liking the wrong picture, so imagine what the discovery of an active dating profile could do. Purely innocent acts can invite skepticism or be a source for insecurity even if they are not material to your relationship in any way. All of this unnecessary drama because we supposedly need these things, as if we've completely lost the ability to find and facilitate relations on our own.

An observation of this brief history shows how we have been conditioned to appropriate excessive merit in relational standards set by these media. At this rate, it seems that we will keep taking this type of social simulation further and further until the importance of who we really are is surpassed by what a digital recreation of your existence can validate. It would seem that forming these artificial dependencies at such a young age has instilled weak relational values in our generation, and there are foreseeable consequences if nothing changes and we build on this rickety foundation. Nowadays, we're too concerned with fitting in and maintaining relations that are not genuinely a source of positivity. A popular doctrine states that you are an accumulation of your five closest acquaintances; however, it may be a little outdated as the number of our social influences have easily increased tenfold in recent times. This being said, it is really closer to your 10 closest acquaintances, if not more. The relevance of this is in examination of the environments we hang out or log in to, and surrounding ourselves with those who are an active source of positivity in attaining our desired relations. With the acceleration of systems as efficient and powerful as these, the process of finding a relationship has been diluted to the point of frantic free-for-all. We have such an abundance of accessibility to view a world that imposes the idealistic lifestyles of others, that our own standards have been mixed up, and we lose touch with the immediate world around us that made us who we are. This has resulted in the now commonplace acts of projecting a fake self-image, like trying to photoshop life, and people are pushing these standards to the next level to actually translate to real life. We're seeing child celebrities who have already undergone copious plastic surgery in their teens, all to maintain an image that is perpetuated by these media. In venerating the celebrity lifestyle, you begin to stress the importance of their achievements over yours and turn to obsession of this media as the solution. Notable remnants of this delusion persist even if you're not one of those people who logs in every day to maintain status among your connections or followers. This media mentality converts to our relational behavior in real life and how we value one another based on our virtual alter egos. As far as relationships go, a dating culture focused on media profiles will be saturated with sociopaths. The resonance of this media prioritized agenda is that our perception of values is skewed from the hopes of finding a knight in shining armor to a "tinderella" or whatever "bae"[6] counterpart that is.

Media standards are potentially obsolete outside the realm of technology mainly due to the fact that vital organic elements of communication are lost. When you put this media on a pedestal, you become subject to implementing dynamics of their behavioral standards in a natural environment. The problem is that real-world experience cannot be gained from these media, so replicating that behavior in person will not always benefit your relations. Media involvement in synthesized representations of being is affecting our communication, but also by extension our most intimate of acts in relationships—sex. We see a booming industry in focused cinematic sex, or pornography, that has taken over the internet. This newfound outlet of instant gratification also promotes the ideals of our hookup culture, in that we find our most instinctual desires met through means of an unnatural environment. In many cases, these porn stars are put through abusive and inhumane set conditions just so we can get our easy kicks. Not only do these types of productions perpetuate the mentality of our obsession with fabricated identity, but they desensitize us to the true meaning of love. Making love should be the most intimate and private form of relational interaction, yet all forms and abominations of this sacred bond can be found on countless websites. The point being here is that real-world acts are more meaningful. If a picture is worth a thousand words, what does that make an emoji? Respectively, emojis are not pictures and they are not worth a thousand words as they are limited to their prefabricated and interchangeable form. This is the nature of creativity in reciprocal relativity as words can be used to paint a picture, and there is very little reciprocity between media standards and natural standards. There is a unique aspect in creative expression seen by the likes of signature brush strokes and linguistic accents, which cannot be replicated through binary code. A system of 1s and 0s is by definition the ultimate form of objectivity, so it is no surprise that there is static interference when we communicate subjective matters through this channel. We can note that artificial intelligence is the ultimate form of this binary structure, but a confined structure in subjectivity is chaotic. Communication, especially in a relational context, can be seen as an art form of creativity making these prefabricated forms of expression over media basically obsolete. This natural communication can be very closely imitated, but not replicated among the confines of whatever form of media expression. This is not a condemnation of these things as they have their moments, but sometimes it pays to go the extra mile. So, instead of using an emoji heart to tell someone you love them, use dialect and the many words we have to choose from in forming a more precise depiction of how you feel. However, getting back to the topic, the dilemma spans beyond communication over the entire realm of connectivity and new challenges to face with regards to relationships. Today comes with the ease of technology that allows you to be a part of someone's life through pressing a button. I believe there was a time when kickstarting a relation meant having the courage to go and get it face-to-face. A first impression by friend request is trumped by a polite introduction hands down every time. This concept that nothing good comes easy reinforces why a solid relationship rarely comes from an app that requires a selfie, a bio sentence, and the ability to swipe left or right. This mentality is in line with the society that gives out trophies for participating; when you do not understand rejection, you have no way of finding where to improve your weaknesses.

Appreciating each other for things that cannot be measured over media will prevent misappropriating objectified media values in our everyday lives. The solution here is in balance. Media can be used to expand our horizons, but they have their limits, and we should not forget what relations mean beyond them. Again, there is no inherent problem with these media; rather, we're too quick to disconnect from real life and login to a profile. The ease of separation is an effect of an abundance of alternatives, and this can result in promoting lack of commitment for fear that something better is around the corner. There is a newly tailored acronym to describe this phobia—FOMO, fear of missing out. The one thing worse than buyer's remorse is simply doing nothing, all the while wondering what could have been, or even worse knowing what you missed. At any given point in time, you can drop what you're doing or who you're with and access an endless amount of alternative options. This potential opportunity instigates the disposable mindset of what we are currently engaged in, like having a sort of relational attention deficit disorder. Not only does this account for a lack of commitment, but in having so many backups, you tend to develop a lack of empathy. This concept is in line with the perspective that a person represented over media is just one among a countless number of profiles instead of a fellow human being. Lack of recognition for individual value leads to mistreatment of relationships because they can be discarded just as easily as they are attained. The reason for this is due to the inverse attributes of quantity and quality. It is much harder to maintain the same quality relationship with 1,000 people than it is with 10. In order to deal with this disparity of relational quantities, we have adopted coping mechanisms in their handling. A recent term has been recognized as ghosting, or completely cutting ties without notice and with no intent of returning. Ghosting is a result of having less of an inclination to maintain a relationship due to the endless prospect of others. Coincidentally, this concept is a figurative embodiment of the Snapchat ghost icon, in how our experience simply disappears. This is a perfect example of where our priorities lie, and how we are beginning to deal with our obstacles by simply giving up. It is easy to give up, and therein lies the key to successful relations, because love is not easy. You must be strong to have love for others in the same way as for yourself; adopt this mental stamina for your relations, and they will prosper. What we need to focus on as an evolving society is to work with the technology and newfound connectivity instead of letting it work us. Yes, today there are more text messages and fewer poems than in the Middle Ages, but chivalry will prevail.

Chapter Three
The Millennial Gentleman

Chivalry—Then and Now

Historically speaking, chivalry and the concept of a gentleman have gone hand in hand, but the time is now for our world to redefine this word to be more progressive. Chivalry has conventionally implied an obligation that men have to women involving characteristics that make him a gentleman, and furthermore initiative actions that are effective in the process of romance. Although the basis of relationships calls for mutual interaction, in many cases it has been a man's obligation to suggest commencement of action. This can be seen in classic phrases such as, "May I have this dance?" or "Will you marry me?" Although these are usually things a man would say, there is no reason a woman should not. This is one of the basics of chivalry in stepping up to the plate; however, this does not suggest a heightened aptitude in either sex or gender, as it takes an equal amount of courage in both positions; it's a two-way street. Although these energies are similar in scale, they are not the same type, but luckily for our generation, we have all this new connectivity serving to level this playing field. When used correctly, these networks open many facets to mutual interaction and reciprocal prosperity in relations as they serve to familiarize the opposing sexes with one another. Chivalry is most commonly associated with the picture-perfect knight in shining armor from the medieval times and those depicted from the legends such as King Arthur and his knights of the round table or Charlemagne. You would normally associate these strong type figures of men due to the nature of their actions being associated with a man's role in society. This being said, gender roles are constantly changing, usually in a manner of progress to equality. Whatever the societal consensus tends to be at any given point in time, the fact remains that anyone can act chivalrously at any time. One does not need to be as male dominant or grandiose as the aforementioned historic figures in order to achieve the status of gentleman or being chivalrous. In fact, there is a lesson to be learned by society here in the ways of ostentation and indulgence, as many of us strive to have the most over being the most. Chivalry does not in any way involve the measuring of amount in either physical being or material possessions, rather possessing and portrayal of invaluable immeasurable characteristics that truly define who you are, and more specifically how you love. As previously stated, love is strength, and in order to become chivalrous you must first be able to endure the trials and tribulations that come with relations. Strength comes in all shapes and sizes, and between men and women there are definitively different types of challenges in relations, but love remains the same.

There are distinguishable components that contribute to the different roles men and women play in a relationship, but chivalrous aspects are transposable. Chivalry is a practice that has historically been reserved for gentlemen, but this does not mean there is a limiting factor for the principles involved. A very close parallel is in being ladylike, but as mentioned before, I am not speaking for a woman's expectations as I have not lived them myself. Although obviously the actions are different, the underlying principles between sophisticated men and women are interchangeable. Even with the difference in behavior between the sexes, the idea of maturity and the prerequisites that must be met should be the same. There is a saying that "boys will be boys," but there is no place for boys among men, especially gentlemen. Regardless of boy or girl, regardless of men or women, there are definitive guidelines to meet the standards that constitute chivalry, and they are basic. Once you master the basics, you can work on refining your true psyche. We are all taught basic ideals—no complaints, no excuses, no sign of weakness, but that's just being strong for yourself. In making the distinction between men and gentlemen, the most important difference in being able to exemplify chivalry is in embracing the moments when you can put others before yourself, or loving. At the end of the day, your relationship will be far more meaningful after the trials and tribulations, and in this you will be rewarded by putting another before yourself. Think about the basic reasoning—how sunshine wouldn't feel so good if it wasn't for rain, and joy wouldn't feel so good if it wasn't for pain. The hard work you put in toward your relationship is proportionate to the amount of love shared and ultimately results in more fulfillment. In the modern era, however, we find fulfillment in self-love, encompassing the idea that chivalry has become a taboo practice. The thought that you should not talk about something or withhold urges to act in a certain way is in reality a form of manipulation and furthermore a lie to yourself and to others. The world would not have this problem if people would just speak their minds, no matter the implications; there is no cost too big for speaking the truth even if for a minor setback, it will serve the greater good. Truth and honesty share the qualities of being pure, and this is practically where chivalry was born in the form of poetry through trying to find the truth in subjectivity. Unfortunately, this type of behavior is often shunned and expelled as we inherit more access to and dependence on trends and popular figures, resulting in a basic accordance in our society we can all recognize. It is during this millennial era that we hear the news more than ever, "Chivalry is dead."

The recently notable decline in chivalry is a result of distinct environmental changes within our society comparable to those in the Middle Ages. Considering the age of knighthood and furthermore the fall of original chivalry in medieval times, we see very close parallels within societal privilege and how this translates into our relations. The effect of connectivity extends beyond the direct implications of relational ideals into how we see the world as a whole and ourselves as one in the same. The empowerment of integration with one another in a global network has brought about behaviors that are in line with the connectivity that was implemented in the Middle Ages. The end of knighthood in medieval times was an effect of kings, lords, and dukes losing their rule, and this gave way to the peasant class and common folk finding empowerment in their numbers and connectivity. This liberation give way to an entirely new integration of society, and a vast middle class was formed with the newfound system of mercantilism. Mercantilism meant that people were no longer under ironclad rule of their noble superiors, and they found new outlets in which to profit and carve their own path without limits. This led to a permissive phenomenon of sweeping independence, but without limitation, these outlets led people to lose touch with their humility. At this point of increasing ego, we see the masses in such competition for newfound resources that they actually bring each other down. We can compare this shift in societal empowerment to that of our modern era, not only with the inheritance of commerce and economic power by our media systems, but the power of connectivity in our social behavior. The distribution of personal information between one another can lead to behavior that translates to the consideration of relational opportunity costs between the sponsoring of self versus others. This parallel holds the potential for enormous implications in the world today and where we are headed if we choose to let history repeat itself. With regards to our capabilities, these systematic changes seem to bring out the more egocentric nature in people exasperating an unnecessary and unproductive arms race, not only in wealth but status among people you associate with as a direct effect. This materialistic avarice instead of holistic selflessness in a social context leads to hollow attitudes and superficial interactions. Fortunately, we are in the early stages, and a reflection of historical similarities offers leverage in this pivotal point in time. It is important to not forget what we can do for others at a day and age when there is such potential in what we can do for ourselves.

Chivalry is definitely a volatile idea proven by drastic changes in culture brought along with connectivity, but it always has and always will be alive. Cultural changes are and may continue to endanger chivalry, but it will certainly never be extinct. Chivalry is really just a reflection of the inner psyche that tells us to do the right thing and love one another. However, with this in mind, it is certainly presumable that throughout time people are simply misled as to what chivalry even is. From the era that chivalry was born, it was shaped by different ideologies in its specific cultural surroundings. One of the major influences on that time, seen in knighthood formations such as the Knights Templar, involved the religious affiliation of Christianity. Even today, this religion is not the same, as countless sects have broken off with differentiating ideologies, but most of the basic principles in the scripture remained the same and shaped certain aspects of original chivalry. An example of these basic principles can be seen in the Ten Commandments and how chivalry is a similar code of conduct that was facilitated throughout most of Western Europe. As such, this code of chivalry is depicted in detail with records from England all the way to France. One of the first of these records includes over 4,000 lines of poetry, "The Song of Roland" circa 1100.[1] These writings shed light on the vows a knight would take, living for the glory of God with valor in faith, protecting the defenseless, and having utmost respect for women. This gave way to many *chansons des gestes*, which were songs and tales praising knights' noble gestures. These stories were well known far and wide, giving due credit to feats and events that were respected in these times. Albeit some of these tales were embellished, the point is that they still revered the actions and conduct. A popular fourteenth century manuscript offers a closer speculation of a knight's demeanor in Geoffrey Chaucer's *The Canterbury Tales*.[2] The knight portrayed in this tale embodies a virtue and composed stature that is expected. He remains indifferent to the quarrels of others, and after receiving praise for his prestige in battle, he remains humble. Most importantly, even as the most esteemed member of the party, he treats the others with respect and doesn't abuse inherent power. Shortly after this work, a list of virtues were drawn up in the fourteenth century by the Duke of Burgundy stating 12 chivalrous attributes: faith, charity, justice, sagacity, prudence, temperance, resolution, truth, liberality, diligence, hope, and valor. These words all have association to a central concept of conscience in honor. Much later, circa the eighteenth century, prominent writings of Léon Gautier in *La Chevalerie*[3] give extensive detail to the concept of chivalry itself, among other things listing the ten commandments of chivalry. The effect of these records pertaining to chivalry served in the years to follow as a basis for what we currently refer to as "the gentleman."

Millennial Knights

With just a basic knowledge of the emergence of chivalry, one can clearly see that its effects on culture have been aligned with valued principles of its society. This speaks for the general characteristics of a gentleman, and of course human nature mixed with different times will produce unique ideologies, but some things never change. One of those things is the main principle of being a gentleman, in that it cannot be inherited and requires constant attention attaining refinement, never perfection. As far as our current society is concerned, it doesn't take all that much to be a man, which has been boiled down to fundamentals of strength and simplified to doing things "like a man." Even so, one of the main differences between men and gentlemen is involved with your conscience. A gentleman knows to treat others as equals, especially in the time when it is necessary to put another before himself. Stepping up to the plate and initiation are part of chivalry in doing the right thing when a significant situation is on the line. This plays a large part in our priorities with relations and how we must answer the call and seize the moment. It's like the concept of a sinking ship, and how life rafts should be first provided to women and children. This initiative as a concept is not exclusive to gentlemen, and it does not suggest any sort of heightened significance in the process of relations, as that would be like saying the process of driving a car is reliant upon opening the door. Succeeding in driving, like relationships, takes constant attention and multitasking in a dynamic environment. There are certain customs that have been given to men such as a man's obligation to propose marriage and a woman's obligation to take his name, but these things are rooted in ceremonial tradition and hold an insignificant fraction of relational functionality. For the most part, chivalry is encompassed by principles of functionality, and it is important to note that men and women both take part in this concept, simply supporting the tasks we are better equipped to handle. An example of this goes back to how between men and women, men are genetically better equipped for physical tasks, which is why they are expected to do them. Throughout different cultures, there are different concepts of expected behavior and what is considered honorable, so it's imperative to approach these matters over global connectivity with an open mind and heart. This proves that chivalry and equality should be harmonious, not subservient or dominant in holding certain obligations derived from societal and biological *accouterments*. A woman can be treated as an equal and be courted at the same time, and she can treat a man in the same way by the transcendence of chivalry. With regards to media, our current societal structure has a way of emphasizing certain points out of context and concentrating on the manipulative side of them for the sake of a corporate agenda. This is why it is important to not let media control our societal structure, and instead implement those timeless cultural values over the media.

In comparison to millennial gentlemen, the Middle Ages of Western European society provided its own embodiment through the code of chivalry, the knight. In terms of chivalry, these individuals stand out among Old English kingdoms for their noble acts, some of which had even become legend among the likes of King Arthur and his knights of the round table. But what does it mean to be a knight, and what is their place in the modern world? As previously mentioned by example of Joan of Arc, this figure is not limited to one gender, or for that matter any form of limiting factor. What qualifies a person for this position is only validated by the strength of support they have gained through merit in their good deeds performed for the community, which today is expanding to the entire world. A knight's quest for glory is guided by the heart with the force of love being the impetus for their actions, and rightfully so, as nothing could be more powerful. We see how knights, although a beacon of the people standing to serve the greater good as natural leaders, also have a personal dependent nature. The love for their people drives them to battle, and true love for their companion is a constant reminder to behave in a noble manner of relation. We've heard the classic tales and bedtime stories of a single knight on an impossible quest, braving even the treacherous fires of a dragon, all to rescue their one true love. Now, this being said, it is the topic at hand as to where a knight would fit in with the millennial era. We may no longer have quests to undertake with dragons and evil wizards to conquer, but there is still a need for this type of behavior to be exemplified in chivalry. In modern terms, these things can be translated to coincide with current culture and what constitutes our relations over global networks and our personal media. Now with all of this connectivity and potential, modern knights must have the same discipline to coincide with our current and notably stronger capabilities. In order for a knight to be sufficiently suited with the tools in the perpetual task of bettering their community and relations, whether personal or diplomatic affairs, they must first inherit a mindset to manifest the following figurative attributes. For starters, a knight has the proper tools to coincide with experience in functionality, using capable means for the prosperity of their community, and their loved ones. In general, a knight requires only three accessories: a sword, a shield, and a horse. Although these items would be literally equipped, they are also symbolic for the balance between offensive and defensive action along with the harmony of that which drives them.

The sword, shield, and horse are the three prerequisites that literally and figuratively prepare a knight for engagement with their obligations and relations. With our current capabilities in connectivity and communication, an old phrase holds more relevance than ever: The pen is mightier than the sword. With regards to our relations, we are currently in possession of a countless number of channels with which to reach out to one another. All it takes to extend a hand or heart to anyone you could think of is to pick up the phone. Although the sword proves a key component to the forward and charging aspects of the knight, this does not justify all forms of engagement so to speak. When Richard the Lionheart introduced the crossbow to the field of war, great triumph followed in many successful battles; however, historic chroniclers of the time described the weapon as unchivalrous. Likewise, use of our current capabilities requires honorable actions, meaning "late-night booty calls" and "creepy dick pics" are not always the best initiative. The shield is a symbol of protection and defense, and it follows the previous figurative analogy for discretion and choosing your actions wisely to abstain from damage. The shield is the instrument of restraint, where a knight must act in imminent regress for the sake of future progress. A widely accepted principle of competition is that defense wins championships. This goes hand in hand with the saying that the best offense is a good defense, and this is due to the concept of patience. Oftentimes, we are faced with the option of striking, but it would better serve the purpose in the long run to endure hardship and absorb energy until the opportune moment. Although this is easily understood, it is one of the more difficult concepts to embody when presented with hardship, so it is expected of those who are capable of chivalry to stand up for what is right in defending truth and honor. The final requirement for a knight is his noble steed, and although it is this separate entity that drives him to engage in his affairs, their movements are one in the same. The horse is a representation of the knight in how it is taken care of and trained, but it shows more than cantering and trotting about in show. The horse embodies the idea of consciousness and holds a great relevance over modern connectivity, being an essential force of leverage in its dynamic and advantageous nature in power. Back in medieval times, it was considered a prestigious luxury to own a horse, and as such, this privilege came with expectation of responsibility. However, as powerful of an engine as it is, it could also serve to be a detriment depending on the rider's discipline. Much like the inherent clout of our modern capabilities, it takes a level of skill to conduct such potential. With this power, you can make or break your reputation in establishing an alternate persona, and likewise a horse can easily buck you off when mishandled, or it can lead you to victory when rightly commanded. With knowledge of these effects, a person has the figurative mindset of a knight and will be more qualified to execute chivalrous deeds.

As for the place of knights in our millennium, the embodiment of chivalry can be traced back to our earliest records and has been forecast in future depictions. The principle is truly unrestricted by any geographic or cultural boundaries, observable anywhere and anytime throughout history and into the future. Keeping in mind that we have not had written language for more than a few thousand years, and even so we see chivalry in these cultures. One of the ancient writing systems with calligraphy in Asia offers a close parallel to the knight and chivalry. The samurai, a sort of knight by design, is the Japanese warrior who adheres to a strict code of honor—so much so that failure or dishonor would often result in the taking of one's own life in *seppuku*. This ultimate sacrifice goes to show that there is more than the commonly perceived brute nature of warriors, and it is ironically explained with love. The literal meaning of the word *samurai* translates as "to serve," which further shows the similarity of obligation between their knight counterpart, which we know to be a symbol of chivalry. This cross-reference example clearly shows a separation of culture, geography, and time between the Middle Ages and written recognition of chivalry, yet the concept remains. This proves that the idea is just a basic part of human consciousness in strength and will not soon come to pass. As far as future projection goes, cinematography icon George Lucas set out to animate the likes of chivalry in space with his portrayal in the creation of Jedi Knights in his famous movie series *Star Wars*. An excerpt from when the young knight Anakin Skywalker begins to fall in love explains that a Jedi's duty is actually reliant on the basis of love. While Anakin shares a meal with the woman of his dreams, Padme, she mentions her thoughts of swearing a lifelong oath to the Jedi Order. "It must be difficult having sworn your life to the Jedi," she says. "Not being able to visit the places you like, or do the things you like."

Skywalker cuts in: "Or be with the people that I love."

She responds in question: "Are you allowed to love? I thought that was forbidden for a Jedi."

He then responds: "Attachment is forbidden, possession is forbidden. Compassion though, which I would define as unconditional love, is essential to a Jedi's life. So, you might say that we are encouraged to love."[4]

This is the basis of these principles in that unconditional love is the source from which rich and prolific lives are inspired. When embodying the true essence of chivalry in your day-to-day persona instead of isolating actions to love a certain way, your energy is more potent and you inevitably attract what you are looking for without having to look at all. The best way to project and utilize your chivalry is through subtle nuances and things that are not expected of you—the details of demeanor, which eventually serve as catalysts for much bigger things. Regardless of the magnitude, these gestures are undoubtedly more powerful when the intent is genuine, and the impetus is true love.

There are countless acts of chivalry at your disposal, but what really makes a gentleman is the structure behind his character traits that herald his actions. There are case-to-case considerations in acts of love where exact repetition will provide different results, reinforcing the idea that love is a subjective concept. However, chivalry is not fully encompassed by the concept of love, and this can be seen by the fact that chivalry holds objective aspects as well. Chivalry calls for measurable features such as having a code to live by like that of knights or any other respectable affiliation because they stand by what they claim. This will prove to serve as a stronger catalyst in relational attraction and speaks volumes about who you truly are when no one's looking. Likewise, contrary to popular belief, one "like" does not equal one prayer—even if it did, there would often be a better way to incite change. Actions speak louder than words, but adhering to a personal code within yourself is the ultimate compilation of actions as it is the mindset guiding them, which creates an undeniable aura and demeanor. This being said, it is important to reiterate to my generation that a successful gentleman or lady not only understands these principles, but also tailors them to the wants and needs of an individual instead of conforming to what society believes to be currently popular. A noteworthy charisma is based on honorable and influential personality traits, and the mentalities which contribute to the positivity and productivity of relations beyond your own. There are countless variations in which one can identify with their idea of having success in loving relations, but the base principles of chivalry are what seem to be in a state of decline within our generation. This concept of a moral code is nothing new: in fact, we can observe writings that trace back to the early ages in defining chivalry. As depicted in *La Chevalerie* by Gautier, we see 12 traits described at great lengths depicting the concept of chivalry at that time. Of course, times are always changing, and so people's attitudes change as well, but this is where we see a distinction of codes in that they are set in stone. For the sake of my given audience with regards to our society and generation, I have adopted and retrofitted Gautier's 12 traits. Although loving traits are countless by subjective nature, a code of chivalry can be boiled down to basic criteria which are accompanied by their own rhetorical benefaction. Listed in no particular order of importance these are the five noble vestiges that should be instilled to a code in the modern gentleman: integrity, intention, intelligence, intuition, and initiative.

Five Noble Vestiges

Integrity is the standard for which we can determine the authenticity of character and honor, and it coincides with a rhetorical affiliation to ethos. Ethos translates to credibility, and "A man is only as good as his word." The importance of having integrity is a fundamental characteristic of chivalry and is emphasized by this saying of what is required of a gentleman. If you make a promise to anyone, no matter what it is, you must go to the ends of the universe to fulfill that promise, *to the T.* Ironically and coincidentally enough, this phrase is derived from a play written by Francis Beaumont in 1607, *The Woman Hater*. In this play, it was said "I'll quote him to the tittle."[5] The word *tittle* is derived from the Medieval Latin word, *Titulus*, and is defined as an inscription stroke, diacritical mark, or serif accent. It emphasizes the importance of the most acute details, therein being handwriting diligence, but more so applicable in a broader sense of following through with what you set out for in giving your word. The objective of having integrity *to the T,* is to do precisely what you have implied with utmost care for not only your self-dignity, but in respect for the recipient's expectations. In being able to walk the talk, you retain substance in your actions being able to execute what you set out to accomplish. Without this substance, your actions are meaningless and potentially dishonorable if it is a lie. There is truth in integrity, and this honesty extends beyond keeping your word. Today, we have the privilege of countless forms of communication through media, so responsibility should be upheld through the smallest of implications. Having any sort of privilege will come with a responsibility of inherent expectations, so what you choose to do with this builds your reputation in who you truly are. Your character is based upon the strength of what you can produce, and if you advertise a false persona, it will indirectly stifle your own fulfillment. In an era of infinite accessibility, imagine the implications your words and actions have on your integrity, as everyone can potentially see your every move that has been broadcast. We have a very large margin of potential to manipulate reality through various forms of expression, and in many cases the truth is indistinguishable, so the burden falls on you to uphold integrity. This brings us to our last point of integrity, which is being able to practice what you preach. There is no honor in propagating viewpoints; if you are not willing to inherit them, the substance and value behind them is nonexistent. With regards to subjectivity aimed more specifically toward our relationships, these empty promises will undoubtedly catch up with you in some form of karma and come back around. This leads to the next noble vestige with projection of integrity and having truthful intention.

Intention is a reflection of conscience where your priorities truly lie, and this brings us to the second aspect of rhetorical influence of kairos. Kairos translates to time, and timing is everything. Everything worthwhile will require maintenance, and *maintenant* is literally the French word for now. Without sounding too cliché, you must seize the now because, again, timing is everything. Einstein theorized time and space to be relative, and this is beautiful because it is scientific and objective by nature, but the principle applies perfectly to a subjective theoretical application. From an equational standpoint in this regard, being a part of this millennial time, and the potential in our connectivity space, the only variable remaining is personal intention. You have the capacity to manipulate space and time if you have purpose and this is what rewards meaning to your relations. The feature that fuels the fire inside us is purpose, and the overall drive and mentality fueled by a passion to live and fulfill your life. This is part of being positive and prosperous, to live for a cause greater than yourself and pass on a lasting legacy is the most important implication of purpose. It does not matter what your personal reasons or how you accomplish whatever it is, only that you try wholeheartedly. Either big or small, you must continue to do the things that will yield results you desire; anything that does not meet your standards is essentially pointless. Chivalry makes no excuses and has no room for indecision. To achieve this, we must move with confidence and intent, while accepting that if we make a mistake, the fault is our own. There has recently been a decline in discipline here, leading us to be hesitant in pulling out before things get too serious. This concept is nothing new, but has certainly been popularized with the likes of the Facebook relationship status indicator "It's complicated," or being on the fence between single and in a relationship. This type of acceptance is just another form of complacency and is nothing noble. Unless lying on their deathbed, if someone ever feels they have achieved everything they set out to do in life—whether it be with relations, working trade, or any other category—that is the precise moment of failure. It is the age-old paradox that perfection can never be achieved but is always to be strived for. Without hitting the end of complacency, it is important to control your fervor as to not push and exceed your limits, simply by understanding balance. Without this understanding, passion can become addiction, and like scratching an itch too heavily, you will only provide temporary relief while exacerbating the ailment. This ties into the next attribute, in giving prudent thought as to where your intentions truly lie and why they do.

Intelligence is the application of competence that extends beyond the limits of practical knowledge, and brings us to the rhetorical aspect of logos. Logos translates to logic, and in a time of unbound access to information this concept could not hold more relevance. What bears the most significance here is finding the distinction between wisdom and knowledge, and that an understanding does not have to serve a specific utility, such as feeling. The well-known quote "Everybody is a genius, but if you judge a fish by its ability to climb a tree, it will live its whole life believing it is stupid," displays this concept perfectly by the perceptive quote on wisdom.[6] This goes to show that we all have the capacity for knowledge, but we must find our own specialties without allowing others to do it for us, and this relates to wisdom. This is another aspect that we will delve into with the vestige of initiative, meaning that no matter how smart or capable you are, you need to act on it or it is useless. The utility of intelligence is in decoding signs, as they can often be misleading, but are usually there for a reason. With the amount of connectivity offered today, it is very easy to interpret the signals of others based on quantifiable features. Instead, it would be wiser to gauge utility of external signs based on your previous self and determine the merit of your utility upon your own scale of satisfaction. Just because a sign says to do something, does not mean you need to do that, and likewise you should not judge others' expression on a surface level. It takes a very open-minded understanding of success to know that you should never pass judgment on another person, but in this day and age, we have become all too accustomed to this. Most of this media is in place for the sole purpose of speculation, so it would be wise to use this in a constructive manner. This same scale of knowledge in one another can be applied to judging relations, that they are all unique and should not be compared with one another. You may find that in some cases you click with people very quickly, and with others it takes more time, but the energy invested does not necessarily speak to the strength of the relationship. With regards to chivalry, a wise person knows that it takes a certain perception of all surrounding factors to finesse the situation and project properly toward building relationships. This type of wisdom is a perfect transition to finding utility in your feelings, and how real chivalry actually savors the opportunity to emote a mindset guided by the heart. Very closely related to wisdom with regards to relationships is the concept of listening to your heart over your brain.

Intuition is the extent to which you appropriate emotional experience into a directive, bringing together the last point of rhetorical analysis in pathos. Pathos translates to emotion, which highlights the importance of accepting things without trying to measure every single contributing factor. This is a testament to the folly of objectivity in relations because as much as you learn about someone, you will never truly understand everything about them. Intuition is the voice of reason derived from your heart instead of your mind, and in many situations, it is the only voice to be trusted. This is a major part of why these media can misconstrue reality because subjectivity and emotion cannot be felt properly over their confines. There is a multifaceted spectrum of energy that we are able to see in one another in person, and this is constantly changing dependent on our and our subject's feelings. However, this lightshow is only visible through a pinhole scope when looking from the perspective of our media resources. The full scope of one's true being is something that cannot be noted in a bio description with a character limit. In other words, what really makes a person's character needs to be felt in person as to make an intuitive deduction. This is a critical key point that needs to be understood by our generation in that the things we are looking for may not be discoverable over these millennial media platforms. The utility of intuition is the reason the conceptualized media scales of energy were discussed in the first chapter. As comprehensive as these media systems appear on a surface level of sharing information, there is so much left unanswered in terms of raw emotion. This has led to a massive epidemic of confusion as we try to understand finer points of what makes each other tick, and it can be limiting to our relations. As popular as it has become to endorse the idea of relationship goals as we see two celebrities in matching sneakers, these depictions are actually almost void of any substantial content. We need to start thinking outside of the box, herein perfectly embodied by the shape of a cell phone or computer, and in doing so, we will adopt a thought process that will far surpass what was once offered to our objective understanding. This does not mean you need to be an outcast and delete all of your profiles, as these things are also a part of what makes us unique. However, it is important to not take on a sheep-like mentality in blindly following the views and tendencies of the masses when it comes to the intuitive void in activity with these media. This brings us to the last point of our five vestiges, in having the strength to make your own way in finding solutions that serve to be a positive relational catalyst.

Initiative is the fifth and final noble vestige of leadership, and it encompasses all rhetorical appeals proving the whole to be greater than the sum of its parts. This could not be more appropriate as appeals to rhetoric were compiled by one of the most influential leaders to ever live— teacher and philosopher, Aristotle. Teaching others is a powerful form of leadership as it is not directly imposed control, rather indirectly enabling others to be their own leaders. What makes a great leader is not their power, but the ability to empower others. Leadership holds great importance in this Millennial era of connectivity due to the nature of how we choose to project our desired persona over various profiles and contributions to others "following" our stories. Unlike the other four noble vestiges, initiative is an action instead of a description. It entails the resolve to achieve what you expect of yourself, without conforming to what others think is cool or what you should do. Individual power is knowing that your value is not determined or inherited by another, and you alone have control over your life. Your birth is the epicenter of a circle representing space and time, everyone starts off at the same point, and the more you move outward, the more space you have to carve your own path into history with that of your own personal timeline. The decision is yours as to whether you want to travel down a familiar line or distinguish yourself. For the most part, unsuccessful people become lazy and travel less distance away from that epicenter. Even worse, they may even remain a part of another person or group of people's timeline, with achievements that have already been had by traveling the same path. Too often in our generation, people confide to the likes of popular trends for the sake of having more "likes." Social media powerhouses not only perpetuate quantifiable validation, but have implemented a system of calculated hashtag statistics to derive what is trending. Needless to say, an initiative individual does not follow trends; they create them. Epistemic self-efficacy is the derivative of every prolific characteristic, and results of inner desire to be a powerful leader. This concept goes hand in hand with confidence, which is one of the most attractive traits a person can display. Many times today, we are intimidated by others given status by society because of things such as having millions of followers, but you must set your own standards and live by them. A balance of confidence and humility brings about composed, influential, and genuine character in their actions—a natural leader. Selfless leading and taking responsibility for those who you have a vested interest in is no longer noble, but it is majestic.

Nobility to Majesty

These virtuous characteristics cover aspects of rhetorical effectiveness and encompass general fields of nobility, but chivalry is in substantial development. Of these five chivalrous characteristics, none is any more important than the others, and there is a vast collection of other traits not mentioned, which can serve to create your own unique identity. As an individual, it is your obligation to be able to differentiate and define your own chivalry, and in doing so to control what you make for yourself as an effect. In chivalry, your duty involves defining yourself as well as exemplifying it to those who you have direct effect on. Arguably, the most important example of this transference is the duty you owe to your family and more specifically pertaining to your children. In the first chapter, it was mentioned that women are seeing more progressive liberties and equality as they should be, and one of the most important verses to this topic is parenthood and the right to choose. As a nation, the United States is at an approximate 30-year low for unintended pregnancy and is at a historic low for teenage pregnancy. This is due to many factors but planned parenthood should be highlighted, and this is a testament to the fact that we are now doing a better job of planning for our future. We are currently in a time of exiling ignorance with the phenomenal abilities in dissemination of information, so now we should be looking toward the future in our youth. This is the ultimate form of creation and responsibility giving way from nobility to majesty. The king or queen of any given domain, much like a father or mother of their household, is expected to rule with care for those whom they love. The nuclear family is now faced with new challenges that we have just recently begun to see side effects of its interference. For the most part, a millennial childhood was not bothered by the internet or its media, as it was only the turn of the century when this media started rolling in. This being said, today's children were not afforded the peace of mind in its absence and have been growing up with access to certain ends that may be premature. In allowing your five-year-old to have a cell phone, you invite normalization of certain inappropriate attitudes and behaviors. This can also put a great deal of pressure on children to reach the maturity of adulthood where these things are normal, creating stress in who they currently are. It is your job as a parent to engage in your child's relational life, emphasizing importance in family structure and preserving the innocence of childhood until maturity is appropriate. Giving your 10-year-old child makeup and Instagram or toy guns and violent video games is the root of behavioral problems in Western society. These behavioral problems extend beyond societal effects and into everyday relations. In this regard, not only does chivalry play an integral part in the proper creation, but also the raising of a child. In having positive influences, a child is more likely to become a functional and successful member of society and their relations.

Adopting chivalry rewards fulfillment establishing affluent personal relations, contributing to your communal surroundings, and establishing an honorable legacy. Chivalry comes full circle in its positive involvement with a microscale of behavior, a macroscale of societal influence, and inherent value to the history and future of humankind. As far as personal life in the structure of family, we often see a concept of recognition to the "man of the house." Although this chapter pertains to the millennial gentleman, it is imperative to reiterate that family structure depends just as much on women and their chivalry. As a child, you are not always afforded a picture-perfect nuclear family, but if that's the case you quickly learn that family is more than blood. In a time with more ways to be connected than ever before, we are able to understand the limitless nature of our relations. This technology and media systems are meant to bring us together on a social and personal scale, but also to facilitate communities and cultures coming together on a global scale. As a result, we see that being a member of our community has much larger implications. What we're seeing with this influx of connectivity is that the majority populous truly has the power. Again, this influx is synonymous with the 99 percent plus in social classes of medieval times being liberated from monarchy by means of mercantilism and rising above the previous rulers. It is nothing but accents and features that make a king, and the only power they have over you is that which you allow. You have the power to determine your honor, and with this in mind, it is prudent to be your own king or queen. By this, I mean having respect and care for yourself and the lives of those you are accompany to. Chivalry is defined in the strength of maintaining control over yourself and the effect you have on others. By extension, you will gain respect and care for the lives of others as if they had been your own, and your relationships will bear more significance. In this process, you will find that you are changing the world simply by making minor adjustments to your own behavior. This investment in the future is coincidentally the first step in creating a basis for relations that will prove to be more sophisticated and meaningful. It is this point in time with our massive amount of connectivity that we need to once again ignite the idea of chivalry, making relations in our world more civil and refined. On a final note, it is important to reiterate that it is the future we must look to, because we have already undergone the incurrence of these capabilities, and it is now our responsibility as to what future generations will make of them.

Chapter Four
The Millennial Relations

Contemporary Love

The basis of relations is initially in reflection of ourselves, but the dynamic of forming self-perception has changed drastically over the turn of the millennium. As a whole, that which shapes our relations goes beyond the self and psychology, beyond others and sociology, and into an accumulation of other perceivable stimuli involving a third party. As of the millennium, a large part of this third-party stimuli includes what we attain from media. These three factors are what contribute to overall relational experience, but what truly defines our relations is what we make of the free will in our human spirit that we choose to live out in a stream of consciousness. Today, this stream is heavily influenced by our media, which is actually not how you see yourself or how others see you.

The counter to self-image is world-image or how the world sees you, which is a result of interaction you've shared with others and the opinion they have formed of you. A large part of what media consists of is mistaken for world-image, when in fact these are just self-images projected by individuals made to seem otherwise. Examples of this concept include buying Instagram followers or posting a like-for-like proposal in hopes of gaining more clout; simply put, counterfeited world-images. Somewhere in between is a true being, not an image, but rather an actual palpable energy. A proper *elan vital* and spirit can be used to center yourself and find balance in your self-image and world-image. The French term *elan vital* directly translates to "life force," but its implications mean much more. This terminology embodies the most fundamental, pure, and innocent aspect that constitutes our truth. To understand this, take the hypothetical of how being alone in this world would affect your persona, as you would be projecting a world of purity free from influence to your instinctual desires. It is the influence of other things that stimulates your thoughts and even shapes your perspectives, and guides the actions you choose to take. Actions will always speak louder than words, and this is why relations depend upon communication, alacrity, and the basic principle of reciprocal energy exertion through engaged contact. In the media we use, people are subjected to one-way transfer of energy at its peak, where the possibility of emotion being felt by one and not others is commonplace. Feelings are impossible without the implication of costing a certain price, which is why media involves so much static. This would seem to be a complete paradox to the order of free will; however, this is the controlling factor, as without control, emotion would be utter chaos and inevitably impede and self-destruct. Take for instance artwork and brush strokes guided by free will. There is order to create an intentional piece, but taking away the intent and free will dictate that the work will cease to exist. Under this pretense, all things serve a purpose, whether direct or indirect, all actions affect everything else in a chain reaction. Like so, our human interaction can be seen as intertwined to the extent that each person is linked to at least one other. This is theorized in the six degrees of separation, or how each connection is active in holding potential to affect any other person affiliated. The butterfly effect is an explanation of this idea in how a single action can persist to affect all things in existence. This being said, relations between humans are very sensitive to numerous stimuli, but are reliant on truth.

Relations depend on a synergy of your self-image and world-image, because somewhere in between lies truth, which can be derived from true love. Relation in its most basic definition describes the connection and the force, which make that connection possible. There are many types of relations in the world, being subjective like personal relation, and objective like scientific relation. For the sake of the millennial age of chivalry, we are dealing with subjective relations, and more specifically our relationships. Although very similar, what separates a relation and a relationship is a level of intimacy and love implication. The objective meaning of the word *relation* suggests the literal connection between two points, but in subjective terms of relationships, the space connection of the two people is figurative. As we can observe through many forms throughout our world, closer relationships entail a level of responsibility, such as the Secret Service's relation to its president. With regards to affiliation being in love, closer relationships create a stronger connection of trust that permits access to personal matters and feelings. Trust is a derivative of honesty, and provides the strongest sustenance for a bond in relation because there is no space left between the two points for misinterpretation; there is only truth. As previously stated, when defining one's character in their image, we must allow for balance, but what this has become as of the millennium is a mockery of truth. Somewhere along the line not too long ago, image conformity took hold of our priorities, and people stopped thinking for themselves. A related example for intents and purposes of this book is in what we commonly refer to as the perfect image, or "relationship goals." In these times, we see people retweeting Pinterest posts in hopes to achieve these picture-perfect moments, but oftentimes this only happens because it is what we find trendy. Ask yourself if this is truthfully what you want from life, or just what you think others may want from you. This is what we need to reconcile, in finding the balance between self-image and world-image, and most importantly finding what you are truly passionate about. If all you want from a relationship is to have a steady amount of external attention, well to each their own—but there's more to love. Defining love is difficult because as a whole it is different for everyone, but there are certain truths that may be identified as its constituency.

Despite how much relational experience you have, you will never understand the true nature of love until you experience the strength of its boundless effect. Love—as the saying goes—takes a second to say, hours to explain, and a lifetime to prove. In the world today, we try to measure love and we even see the most common representation, a heart, set forth in the form of recorded "likes." Needless to say, liking something is much different than loving, so this representation devalues the symbol and the notion, and we are off base for trying to enumerate it. The heart has been a beacon of love for quite some time, and people talk about it as if it had a separate conscience than the mind. Listen to your heart; the heart is the life source of our body and reacts to our feelings in its rhythmic nature, especially when in love. Love is the recognition and acceptance of another person that has no formula or reason. It comes as a result of a focal point, whether it be love at first sight or a gradual realization, wherein you are positive of its existence beyond reasonable doubt. Even though this recognition cannot be defined, it is most certainly felt and understood on a subjective level. The bond is at its strongest when acknowledged by both parties in harmony, and consideration for one another is balanced in energy exerted toward the greater good of the relationship. This is a major factor contributing to the difficulty in finding and facilitating love over media, in that there is so much room for misinterpretation. Loving recognition represents your ideal counterpart that compliments your potential true image, so when these images are distorted, the feeling can be lost. It is finding admiration of the things that you value most and want for your life, but have not yet learned or are otherwise incapable of alone. It is having consideration to the respect that you value your lives as equal, in acting selfless through anything and everything as if it were directly beneficial to yourself. In conjunction with the emphasis on sharing these desires, this togetherness will lead you to become one in both wants and needs. Of course, we all have unique wants and needs, and they get even more specific when combined in a relationship. This is why many people have a different definition of love, and a testament to why there are so many types of love — true love, fake love, tough love, gentle love, brotherly love, motherly love, puppy love, old love, and so on. This is why a feeling such as love is a mentality and a choice, which can be attributed to any number of unique internal or external sources of stimuli. The ability to love is ingrained in everyone because it is of the mind's purest intent to cherish and nourish the self and one another, but it is our willpower that decides whether we embrace or deny it. Love is the strength of our free will, and hate is its weakness. Love is a caliber for which the strength of your bond in a relationship is measured, so it is important to express it in the most effective channels.

Your relational strength has to be directed toward a vision to proliferate the care you have for another; in other words, be the change you would like to see. Whether you believe something to be a feeling, a mentality, a vibe, or any other form of manifestation, it is undeniable that these are all channels of energy. A theory accepted in mainstream science is the preservation of energy, being that nothing is ever created or deleted from existence but rather transformed into another form or forms of equal energy. Similarly, as it is involved with our connectivity, personal information trails are getting harder and harder to sweep away, increasingly so as our digital fingerprint gets more detailed. What were once file cabinets have been turned into data servers, and personal information is stored in clouds—we even see virtual coins holding a standard of currency among intangible vaults. Alongside these technical implications, we're incurring behavioral transference of energy that is relayed from our media. We are beginning to care about parts of our relations that mostly pertain to the perceived success among these media systems instead of the interpersonal skills and attributes we hold outside of them. There are extreme behavioral tendencies that you can adopt from just your cell phone, and they are beginning to spill into our everyday lives. As a result of the way of life we have become accustomed to and the environmental factors surrounding our generation, millennials tend to be predisposed to social and relational shortcomings. For the better part of our lives, we have been conditioned to believe that many things in these technological capabilities are the norm when in fact they are quite unique, which can ultimately create a slight disconnect from reality. We can even see this acknowledgment of differentiation with the increasing trend over the internet to refer to the outside world of reality with the acronym IRL—in real life. It will not prove to be beneficial in emphasizing the importance regarding accolades among these media in place of the experience in real life as we often care about the wrong things. We need to understand that although our capabilities are made special, we are not, and we need to go through the same motions of life's trials and tribulations without the expectation of special treatment. The strength of our character dictates what we are able to create of our relations, and this has shown itself as being the foremost threat in being the source of our relational deficiencies. The relational process is a never-ending endeavor to find understanding for one another in our shared strength in love to work with each other alongside our new environment. An effective practice to balance strength is to focus on weaknesses as to become well-rounded in character. In order to be the change you want to see, you must work on your weaknesses, so I have condensed millennial deficiencies into a three-step program, or the *Three Ps*.

The Three Ps

Be passionate. Millennials tend to feel entitled to our personal aspirations and expect that relational desires will just come to us, but this is merely complacency. Relationships thrive off passion so this is critical, but the effects of these millennial environmental factors extend into our everyday lives in other areas as well. A result of having an abundance of information, technology is seen in how we feel that hard work should be auxiliary to smart work because we are so much more advanced, but there is much more to life you can't obtain on the fast track and online. Real-world relations don't really care about how advanced or popular you are on these media—successful relations take hard work and real-life trial and error. Your relations will inevitably go through ups and downs, but it's the strength and support you are willing to give that fuels the process of building a strong foundation. You need to be strong in order to remain positive and focus on growth, which can be seen as one of the prolific benefits to our capabilities as we are able to grow our social and relational connections facilitated by these media. The fact that we can more easily keep track of one another serves to strengthen the bonds we share and set precedent for future relations. Although there will be many things we can accomplish over media, we still need interpersonal skills, and these must be earned. An example of this beyond a relational point is in how we give participation trophies, which actually devalue accomplishment. This concept is reflected in relations when we log on to a place where we can compare superficial and oftentimes worthless achievements such as a "like." It is an attribute of these media to make it difficult in assigning unique value to this type of validation because we concentrate on numbers instead of the people behind them. This being said, these achievements may not provide value outside their confines, and may even prove to be detrimental. In its respect, you can note that dopamine is released every time you get this type of validation and feel special, which is exactly what these media systems are designed to do. Dopamine is also released when we smoke, drink, and gamble, yet there is no age restriction on any of these media outlets. Using media as a source of validation in your relations may prove to be addicting, and its implications can be similar to that of smoking, drinking, and gambling. In moderation, these things are to be enjoyed, but abuse can lead to a relational disaster. The more we depend on these media for relational gratification, the weaker our real life relational skills will come to be. This is why we must be passionate in situational context outside of these media parameters to build courage and face real life feelings of comfort, awkwardness, happiness, sadness, and everything in between. Sure, the spoils of these media are much more easily attained, but the spoils of real-world experience are unparalleled in value. This endless stream of validation at our fingertips leads to entitlement in relational standards by these media, but can also lead to a disposition of expecting immediate results and a lack of our first attribute, passion.

Be Patient. Millennials have been presented with capabilities that gauge proficiency with speed, which can lead to a dependence on instant gratification. We have been given the tools that make our lives much faster paced; in fact, we can go online and order almost anything imaginable to our doorstep by overnight delivery. Not only does this concept contribute to amplified consumer behavior, but it also has social ramifications in how we are able to attain immediate relational feedback from media. At any given time, we can choose to pick up or drop the attention of people we choose to engage with without a single moment's notice. This can lead to a disposition to hesitance, lacking of commitment, and ultimately stifles deeper relation and love. Of course, the possibility of love at first sight or instant connection is plausible, but our capabilities do not make these things more commonplace, and they should not take away from traditional methods in gradual and stable building. In fact, a simple argument can be made in the "easy come, easy go" case. We need to learn to give things a chance and to not give up on our potential so easily. Here, we see that patience is about more than waiting; it is setting yourself up for when the timing is right. In terms of our relations, there are people who are single with children, who are married with no children, who are in a relationship and love another person, who love each other but not in a relationship, who are waiting to love, and waiting to be loved. Despite your current relational status in life, you are not ahead or behind. You are exactly where you need to be at this time, so be patient and calm yourself. This takes discipline with an understanding that failure is inevitable, but what matters is how we deal with failure in keeping an optimistic view and making steady progress. In coercion with the first step of passion, we must have a vision and follow through no matter how difficult or how timely the cost. A relation is truly a never-ending process with changing dynamics, so in order to succeed you must commit to the cause and keep an open mind. Much like the "swipe" dating app habitualization, we have developed "an on to the next" mentality, but all good things come in due time. Patience will also prove to be a combatant for any irritability at the dissonance of relational trials and will bring you to appreciate the fruits of your labor. Through patience, you will find that time and energy spent in endeavors will no longer cost a price, but will rather will act as a growing agent in part of the relational process. As such, you will come to see the bigger picture in how timely things strengthen your endurance while learning to enjoy the moment without trying to achieve a goal or reach some type of finish line, as life is truly about the journey and not the destination. A common effect of having a non-stop stream of instant connectivity that is consuming our attention is in how we fail to appreciate the moments where we can stop and smell the roses. Appreciation is the currency of relations, and this empathy measures your relational wealth — savoring every part of your being together during the present.

Be present. The millennial environment has such a vast amount of auxiliary engagement that some of us have developed relational attention deficit disorder. We are constantly bombarded with alerts and notifications as we shift back and forth between media in the hopes of staying connected and up-to-date. However, being up-to-date in this context is not the present form of being that will bring appreciation for your relations. In fact, it is often beneficial to disconnect, finding appreciation for your immediate surroundings. Appreciation should be valued at all times, especially in the present, as a relationship thrives off recognition and positive feedback between one another. The importance of presence revolves around the idea that there is a difference between awareness and actually caring about what's going on. This being said, our media provides a great channel for sharing feedback with the ones you care for, but there is a time and place for everything and, if it interferes with your present company, it may hinder those relations. In a world where we can figuratively be in multiple places at once, this holds utmost significance, and the key is simply balance. These media can be a great tool in having a social presence among a global community, but you must not forget the importance of contributing to the people around you who are directly a part of your life. There is a notable point in developing interpersonal skills derived from presence that will lead to having a positive impact on your local community, whereas focusing your energy on media will mainly result in being internet famous. Relationships depend on real-life engagements and application of these interpersonal skills, so behavior that impedes these processes will come back to haunt your relation. A classic example of this disruptive behavior in engagement is cutting a person off when they are trying to communicate, commonly referred to as snubbing. In the coming of the millennium, we have seen an evolution of this type of disruption in "Phubbing"—phone snubbing. This is basically the same concept of taking part in real-time face-to-face interaction, but abruptly switching attention to your phone in an untimely and impolite manner. Not only is the immediate act off-putting, but it can often be contagious by encouraging your present company to pick up their phone in joining, or rather leaving you. Being that our perception of relations is subjective, you certainly reserve the right to a preference of immediate presence and practicing love deeply or vicarious presence and practicing love widely. It is often beneficial to have balance in these behavioral approaches, but depending on the relations you are looking for, it is useful to define the characteristics of these two ends in terms of long-term or short-term relations.

Millennial Disposition

There are various attributes in forming relations, so we can best serve their interests by abridging and assigning behaviors to short-term or long-term input. The aforementioned parallel made between the two involves the invest-ment model of loving widely or loving deeply, whereas short-term relations are more common among those who love widely, and long-term relations are more common among those who love deeply. We can identify the traits of these two in a condensation of generalized behavior and which type of relation they contribute to. In short-term relations, we see practice of mostly passionate love, more often containing traits of lust, desire, fervor, and so on. In long-term relations, we see practice of mostly com-passionate love, more often containing traits of sentiment, affection, empathy, and so on. Of course, there are always 50 shades of gray where you experience both at different levels; herein lies the importance of your ability to identify the two polar ends as to find a balance that fits your perso-na. There is a certain portion of inevitable paradox in clas-sifying a relationship where you must notice discrepancies between your personal perception and the collective view of your societal surroundings (self-image vs. world im-age). An informal classification can be observed in sorting with social and dating media's scroll down menu, ranging from "single" to "it's complicated" to "in a relationship." A more formal and officially recognized classification can be seen in government or religious sorting such as "mar-ried," whereby taking place in ceremony is accompanied by legal paperwork. Whether you see your engagement as something simple, such as friends with benefits, or something serious, such as life partners, it should not be for anyone but the two of you to decide. *Labels* are merely assumptions made by self-serving omission of truth, and have no capacity to be set in stone as to be *defined*. With regard to sexual preference and relations, labels are just a generalized categorization based upon inference of intention, while a definition is a distinction of character based upon proof of action. Feelings are susceptible to the pressures that can shape your life if you let them, so it is better to disregard labels and define yourself. In today's era, it is commonplace to seek labels which confirm that we are on a set path, offering some kind of security for our status. A square is a square because of concrete criteria, but love is not the same love for anyone. Regardless of whether your love be facilitated through physical or mental stimulation, promiscuity or fidelity, or any other type of personal criteria, all that matters is that you are honest and at peace with your intuition. Labels impose false duty to follow guidelines set by precedent or societal norms, and a perfect depiction of how they can be toxic to matters of subjectivity is through art. There is a famous painting by René Magritte, in which he depicts a simple pipe, and in the French language it reads, "*Ceci n'est pas une pipe*," or in English, "This is not a pipe." A literal interpretation can be justified because it is a painting and not literally a cylindrical appliance, but figuratively, it gives much more. This being said, the labels of sexuality and relations—girlfriend, fiancé, wife, bi, gay, straight—have multiple implications that completely change the dynamic of what you started with in a natural relationship. There is an infinite number of ways in coming to accept and define a relationship, but the important part is not what it is but rather why it is. In millennium times, we see people forming relationships all because of pressure from labels, which unfortunately affects the way we express ourselves. Today, the equivalent of "shouting it from the rooftop," is merely posting status of relationship change. We need to get back to having pride in who we are, and the first step is in understanding who we are by the relations that we belong to. The easiest way to combat these challenges is finding clarification through classification of relations in the aforementioned short- and long-term intervals, and the contributing factors of the two.

A challenge that millennials face with short-term relations is in their inability to legitimize and follow through in making adjustments toward a meaningful bond. The primary threat to legitimacy in relations with media is in the transparency of feelings with "fake news" reports of emotion. We now tend to use media as a means of deflection from serious engagement with a less labor-intensive substitute of avoidance. It is now easier than ever to project a sideways message to better serve the short-term solutions to relational conflicts. One example of these types of media enablers is to "subtweet" or offer a micro/passive-aggressive post toward the situation at hand. This behavior is counterproductive to the greater good of the relation, which brings up another cause of this short-term inaptitude in a lack of incentive for deeper commitment. When there is a problem with the level of commitment to a cause, the moving parts and mechanics will seem like more of a chore than something to be enjoyed. What seems to be a major contributing factor to this challenge is in how relations can contain a faulty interest based on merit instead of desire for connection. This would be like dating someone rich, famous, or good-looking just to show off the surface levels of their being like a trophy. This practice is often perpetualized by objective qualities in our media, which misplace measurement of human character with their own valued criteria. This type of external media validation takes away from the consideration for the ones directly involved and ultimately promotes short-term relations. Conformity to the purpose of these media instead of the purpose of your relation is backwards thinking and results in disposable relations and failure of valued relations. Another conflict of interest can be based on the level of intimacy you share together, in that there is either too much or too little. Many short-term relations depend on strictly physical relations and purely sexual drive, commonly referred to as fuck buddies. On the other side of the spectrum, the "friend zone" is a figurative place to be when the two parties involved are opposing in their long- and short-term disposition. This millennial generation is battling between these two ends mainly because of the influence of media that encourages and emphasizes sexuality and physical presence. This is the systematic doing of what has created the familiar relational society as being the hookup culture instead of a genuine dating culture. We've been subconsciously and blatantly led to believe that the behavior we hear and see on MTV and other reality TV is normal, and it is acceptable to have a bachelor(ette) variety of simultaneous relations. If you are looking for short-term relations, you are most likely attracted to the qualities that appeal to surface level qualities or instant gratification that can be attained by the likes of these swipe dating apps. It is not to say that this is the right or wrong way to live your life, but it has certainly taken a dominant hold on our millennial culture, and it should not be forgotten that long-term features are also acceptable.

A challenge that millennials face with long-term relations is in how they find alternative fulfillment in temporary stimulation in the place of emotional commitment. This concept extends beyond obsession with instant gratification toward an addiction to a constant dose of feel-good disconnect. Of course, every now and then relational parties should give each other space, but completely disconnecting as to ignore your issues is unhealthy. A part of this can be seen in the case of when we do attain a long-term relation, but find ourselves dissatisfied, and accepting of this depression with unhealthy coping mechanisms. Many times, instead of facing our relational disputes, we simply "block" one another and move on as if nothing happened. This is an unhealthy way of dealing with the result of your actions, and leads to a lack of closure with a looming cloud of anxiety and future relational dysfunction. The other end of this aspect is in the amount of past relational information we have on display, which can affect the future, otherwise known as baggage. With the amount of information technology and connectivity surrounding our lives, it is easy to dwell on the past and bring up irrelevant obstacles. This is not to say that all these media are harmful; in fact, some websites such as eHarmony boast a high turnover of marriage rates among their members. Herein lies a classification of long-term relationships noted by the idea of monogamy, but we tend to be intimidated by this idea, mostly because it is misunderstood. Monogamy is simply adhering to the principle that you will be with one person at a time, but we often misconstrue this as a commitment to matrimony, which is constructed around the idea that you will be with one person for the rest of your life. Millennials have a hard time with this concept and are waiting much longer than our previous generations because of the major investment and risk involved in the unstable times. In this practice, it can be intimidating to look as far ahead as spending the end of your life with one person, because you begin to anticipate negativity down the road that may not even arise. To avoid this perceived dilemma, we often find it easier to resort to media in facilitation of numerous short-term and less emotionally invested relations. As appealing as this immediate solution may be as an alternative, it is textbook redirection. A safer approach to serious relationships is by simple recognition that monogamy can be achieved in a basic agreement of exclusivity without pushing for absolute labels or eventual matrimonial commitment. Matrimony is a constitution as old as humankind, derived from a bond, which two lovers shared in vows and commitments; but it took a whole new form as feudalism emerged in the Middle Ages and resulted in diplomatic marriage for control and power. Today, we see matrimony even further institutionalized as it comes with irrelevant legal implications and random tax deductions. This circles back to the idea that marriage is a label for something that has been taken from its natural environment, and put into classification to be accounted for. Being that these types of labels may have unwanted stipulations, it may be best for you to just ignore their construct until the time is right as to concentrate on building the relatior instead of classifying it.

Monogamy Mechanics

Our personal preference and opinions of short- and long-term intervals may vary, but it's undeniable that all relations entail a lifelong process of refinement. Through this concept, a valuable lesson engrained in chivalry is revealed, which is that perfection should always be strived for, but will never be attained. There is always room for improvement in subjective matters, especially relations, so we must value and embrace the lessons that come. A major faux pas among millennials is a tendency to overlook the significance of their relational participation, brushing it off as flings or casual encounters. It could very well be because we are so used to this hookup culture or maybe because we no longer believe in lifelong relations and soul mates. Whether you believe in the concept of finding a soul mate or not, there is a lifelong impact from the experience of your relations that shapes you. You may never find one soul mate, and even if you do find what you believe is one, you may not succeed in joining or keeping that bond. It may take several serious relationships to get it right, but as the saying goes, "it is better to have loved and lost, than to have never loved at all."[1] This is because love is a teacher to us all, and if you take the right steps in educating yourself then your relations will provide fulfillment regardless of the hardship and heartbreak. We now have a capacity to connect and stay connected, almost to the point that it is unavoidable, so it is important to learn how to maintain relations with a level of functionality regardless of prior engagement. Although today there is a wide variety of media systems that serve as a wonderful tool to facilitate friendships, casual hookups, and serious relationships, it is important to remember that these are still just profiles and that the real world will better equip you with experience and overall relational success. This is not to say it's all work and no play—love can be all sorts of fun, but there should also be a line to draw when you begin to play with another person's emotions. Relations are not a competition; there is no winner or loser, yet we often strategize and contrive false notions in order to gain some delusional upper hand. A tactic such as playing hard to get is an example of this type of classic mistake in the novice stage of flirting, dating, or courting—especially when maintaining a relationship. Forms of manipulation seem to be more commonplace due to the society we live in that encourages said players. This age-old game of cat and mouse promotes a fundamentally flawed perspective of relational morals in being genuine versus fake, and replacing your identity with an alias, which can be speculated as a specialty feature in these media systems. There is no race to be won, so it is important to pace yourself and move the relation forward at a speed that allows you to keep in touch with one another—take it one step at a time. Thinking too many steps ahead can cause you to trip, so fluidity and moving forward at a comfortable pace will prove to be most effective. Many of us go our whole lives through numerous relationships with lots of ones, but not the one. Finding the one doesn't necessarily mean there is only one, as today's global population exceeds 7 billion, meaning chances are there are more than enough people in this world that can make you happy. The only obstacle to happiness is having unrealistic expectations and standards set to levels of delusional perfection—yet another faulty contribution from media image agenda. True love is finding someone who will make this happiness possible despite all of their imperfections; it is unconditional.

Whether unconditional or impartial, in one way or another all relations come to an end, and there is an important aspect of chivalry in the idea of resiliency. Nothing is beautiful because it lasts forever, and at every end is a new beginning. There is a natural balance of life that gives all things inherent value in that they are never truly at a still point, but rather fluctuating back and forth. With regards to relations, it is normal to have been let down, but you should embrace the fact your emotional capacity has expanded, and learn from it to make you stronger in lifting yourself up again. This is why it is important to be able to accept others for their differences and maintain love in your heart. It does not matter if they are a stranger, friend, mate, family, or even in an ambiguity with yourself; it is imperative to find acceptance to maintain a healthy state of consciousness. Fighting without cause is pointless and destructive, so it is reasonable to establish a resolve before it derails into something unnecessary. A helpful tool to align constructive conversation in today's age is to separate from negative influences surrounding you, as many arguments stem from insecurities of the self that are a cause of a non-conducive environment. It is important to properly gauge the relevance that points of the media hold and if the juice is worth the squeeze. That is, if their relation is worth more than the problems that come over time. Accepting another's perspective can be the hardest thing to do, but this is almost always the most effective way to advance in a relation when sincere. Resilience is about enduring the negatives while maintaining a focus on positives, and we have seen accelerated growth in opportunity for building relations. This concept of moving forward and releasing any toxic habits is crucial in a day and age where all things are open to such concentrated speculation over media. A tip: don't reopen past wounds whether in person or in cyberspace; try to keep the present and future as to avoid baggage. Yes, things such as an infidelity should be called to attention as it is noteworthy, but trivial matters such as liking an ex's photo may prove inconsequential to your shared love. Is the root of your problems based on the single occasion of someone liking your partner's profile picture? To this respect, a phrase that can often be misleading is "forgive and forget," as this will only enable the other person to continue making the same mistakes. Everyone makes mistakes, so it is important to learn how to forgive, but part of loving yourself is not allowing the act to continue if a problem persists. Although you should not completely forget anything, it can be a self-manifested obstacle to moving forward if you are constantly looking back. In order for a relationship to grow to its full potential, you must support one another in all of the strength you have to spare, as exertion of strength is rewarded with more strength.

The final stage in being the pinnacle of relationships is the interlacing of families and the prospect of conceiving children for bonding of a legacy. This continuation brings a possibility of life that you are directly bonded to and for whom you hold responsibility as if it were your own. Family is the center of our civilization, in that we have more power together in cooperation and partnership. The general purpose of chivalry is to uphold love and empower one another, and in doing so strengthen your relations and by extension your community. The prospect of bringing together families and/or bringing life to your child presents an opportunity to focus a transference of love and knowledge. By taking your relation to this level, you are not only enjoying all that life has to give, but giving and fortifying a catalyst that generates a direct energy of potential to be the future. The family unit is instrumental to chivalry by endorsing a commitment to real-world experience of what it means to maintain a lifetime relation. During the Middle Ages and the widespread recognition of chivalry, a large part of determining character legacy was recognized in your coat of arms. The family crest was symbolic of the deeds and reputation of your name and the status you had earned as a means of legacy. During the times of our millennium, we have detailed information to account for the actions of ourselves and our family lineage. Our media profiles serve as a form of coat of arms, only much more comprehensive in depicting our relations and the effect we have had on our community. The records continue to be updated ranging from long before this information technology and into the future. The implications of this are enormous for establishing a precedent to who you are and what your ancestors have done for you, as well as what is yet to be considered. Instilling chivalrous tendencies to the ultimate level of relation will secure a legacy that your family can be proud of, and provides a model for others to look up to. This concept is at a high degree of importance in times where chivalrous tendencies have become less common, as to stand out and establish yourself and your relatives as leaders in a movement toward the greater good of our world. Formation of a family is no easy undertaking even without all of the distractions that we face today, so it is important to emphasize love and utilize the best of your relational skills. Again, this is the ultimate form of relation and entails an unsurmounted level of responsibility, so you must be conscientious to focus your energy on not only cherishing one another, but creating a positive environment of love that extends from your family into the lives of people you are able to give to.

Pay It Forward

Ego antithesis is living for something greater than yourself, and chivalry facilitates this by doing that for which you are better equipped for than the next. What if I told you all of your media profiles had been wiped out, and the likes of virtual identity representation would not be back in your lifetime? How would this affect your ego and how you find confidence in yourself in a way that is not for yourself? The hurdle that some socialites among these media are faced with is that outside these platforms and constituent circles, they do not have the same moxie and find it difficult to relate; if you cannot relate, you cannot help. When a person with millions of followers interacts over a media channel, they have certain measurable clout and act accordingly. When surrounded by people who have no interest in or point of reference to an individual's media existence, they are much less comfortable interacting in the same way. Phones, laptops, televisions, and other media channels provide us with a realm of comfort in the synthesized routine nature of their being. This "safe space" is somewhat of a trap, as relations are not meant to be routine and do not grow in a comfort zone—they are fueled by sublime moments of mental expansion and unfamiliar feelings. In today's world, we remain connected to each other and our exes— keeping cell numbers, remaining "Facebook friends," following one another—because this provides a sense of security. When security is the reason, these acts are actually counterintuitive as moving on from something that has run its course will provide incentive to find new and better things, especially when disconnecting from toxic people is often a click away. To the point of chivalry, you are doing toxic people a disservice by remaining a part of their lives if you don't stimulate their weakness. It is a part of chivalry to help those around you be stronger in ways they do not understand or are not yet capable. There is always room for positivity; you just need to find the strength and love in yourself that others may not have shown you. A purpose of relations and life in general is to provide love and strength in areas that the ones you care for are lacking. All things in life come full circle, and the same principle applies to our relations in how reciprocity is essential to your fulfillment and vice versa. This process can be seen as analogous to the French concept of liaisons, originating from the connection between two words, it means to round the texture of the dialect. When you do or feel for someone, they are undeniably drawn to you, so in doing or feeling what they cannot for themselves, you incentivize them to return the same for you. Doing for others what they cannot do for themselves is a core concept of chivalry and positive relations. Your relations can lead to a prolific life in creating relativity among community, friends, and family; and here's the caveat, you don't need media to achieve this.

Your time here is truly about the creation and elevation of life; whether it be aimed toward your relations or the world around you, pay your love forward. This follows a core concept of chivalry, in that those who are capable of upholding good have an obligation to do so, in effect creating a perpetual cycle of support in strength and love for one another. Here, we see an applicable contribution to this aspect of chivalry engrained in a central theme of government within the country in which it was derived, France. *Liberté, Égalité, Fraternité.* The purpose of these three things working together is to show that our lives are dependent upon what we can contribute to one another. We must come to acknowledge the importance of our relations with our fellow brothers and sisters and why we need each other equally to attain a true sense of freedom. There is nothing we can truly achieve alone, and what is would not be a true achievement. The relational power of our time through the media we literally hold in our hands has the potential to disseminate positivity for humankind to the point of transcendence. Through these capabilities, we can learn to understand and appreciate each other in more ways than ever, giving our relations a chance to flourish beyond limits. Appreciation is instrumental to the lifestyle of chivalry, in that we savor the opportunity to put others before ourselves in the prospect of relating. This creates mutual gratitude, a fundamental form of kindness that shows people you care, no matter how big or small the matter. What really matters in showing kindness is not necessarily these things that can be measured, but rather the gesture, being more about what you mean rather than what it actually is. Paying it forward to this sentiment is key, and to optimize levels of genuine sentiment, you must consider the perspective of those you wish to reach. Treat people the way *they* want to be treated. Everyone has different wants and needs, and a misnomer that we are taught since childhood is to treat others how you want to be treated. When people recognize that you have their best interests in mind instead of your own, they begin to allow you an increased amount of trust. Having trust in one another is an integral part in chivalry as it shows people the value in sustaining moral obligation. This circles back to having an obligation to serve those whom you are more capable than, and building upon what we have as to lift each other up. This is the basis of strength in love with our ability to share in both hardship and fortune of our relations. These technological media and systems of connectivity provide an unlimited platform, which allows us to reach out and extend a helping hand to other people around the world and shows us that we are never truly alone. We will always be able to create relations and elevate each other if only we take the first step, and we will always be able to find what we are truly looking for.

Chapter Five
The Millennial Message

The Millennial Perception

Perceiving clearly at its foundation begins with the ability to understand your psyche, enabling you to honestly project your will for others to process. The operative word here is *honest*, and it is the truth that will empower you and your relations because who you truly are sets the tone for your relations. Coming to terms with truth will allow you to answer the question of who you want to be and will give you the bravery to ask others who you can be to them. Truth in a subjective field such as relations is best observed through the nature of philosophy, so we can take note from one of the aforementioned influences in this field, Aristotle. The first condition for the highest form of Aristotelian love is that one loves oneself, and it is this egocentric basis, which enables the extension of sympathy and affection to others. In this application, we find self-reflective contemplation and meditation to be useful practices when searching for what makes you, you, and will lead to psychological strength to project that to others. Being true to yourself will benefit you with inner peace, and just as importantly, in being true to others, you will manifest a world of peace around you. Although the millennium has vastly increased our range of perception, much of what we perceive through our media isn't always an accurate portrayal of truth in reality, as content is often taken completely out of context and even tampered with. There is so much picture beyond the frame, so much message beyond the text, and so much energy beyond the wires. Authenticity is vital to information for the sake of understanding and learning, but technology has also taken hold of information for the sake of leisure. Today, in place of getting outside and soaking in the sun, we stay inside and find warmth in the glow of backlit screens. We not only find this artificial comfort in media, but we even create alternate lives based off preset parameters. We see this most prevalently in video games, which are solely based upon creating a virtual life in a fictitious reality. The problem here is not in how you find enjoyment in these things, but more in what you get out of them. To a relational respect, instead of finding fulfillment in these media, our time invested will yield far more with real-world experience. These energies are more giving than the systematic activities we've created because they are natural and truly interactive to our physical existence. It is said that the key to a successful relationship is communication, and although these systems hold large amounts of potential here, there is a level of communication they cannot provide. The greatest challenge to communication is the perception that it has taken place. This is where media can emit static interference in the entirety of your true being. Relations depend upon identifying truth and accepting that there will always be differences between one another, and embracing them. The relations you build, the foundation on which you choose to build them, and the implications they have on your life are a matter of your sole discretion and how you will be truly known. The beauty of finding yourself is in the achievement that enables you to know what you are looking for externally, and how people are looking in toward you. This is the best way to prepare yourself for a relationship because knowing what you're getting yourself into and how to remain true to yourself when you're in it is the first step. You must be true to yourself about what makes you happy. No matter what the cost is, you must remain confident and zealous in this pursuit, as it is easily the most important thing in finding fulfillment. Many people, even the ones we share feelings with and trust, are capable of putting pressure on us to the point where the life you once pictured for yourself is changed. This is a major characteristic of our media, but in contrast we see that true love will accept you for anything and everything you are and are not. This acceptance will promote clairvoyance which is not objective in nature, but certain parameters should not be disturbed. We often allow interference of our clarity by trivial matters such as whose agenda we fit in to, which photo will receive more likes, what is currently popular, and so on. We need to minimize this external validation to be in acceptance that things are just out of your control, and sometimes there is no right or wrong. In terms of fulfillment, if you are not feeling euphoric happiness with the way you are right now, you need not fret, as this only means there is that much more to look forward to. To the point of your future in life, as long as you keep an optimistic viewpoint and continue to be proactive in making the changes you want, the pieces that complete you will all fall together.

A practice of vicarious perception through adopting another point of view in applying relational standards will help you find chivalry in the nature of balance. A basic understanding of perception will suffice in the ability to identify and love yourself, but when it comes to the ability of identifying others and loving them, we must be stronger. In this era, we have an immense amount of access to detailed personal information through media profiles; however, in conjunction with the saying "never judge a book by its cover" we need not dwell too critically on these types of impressions. This concept goes beyond seeing only a Facebook cover page and basing assumptions on profiles, into the observation of factors that can only be seen with a level of empathy. This means finding a way to let your walls down and giving them a chance to let their walls down as to uncover relational potential instead of what is apparent on surface levels. An example of this can be seen in the difference between looking, seeing, and engaging, which is the attention to detail and the perceptive analysis of how and why things are in relation to your being. The power of engagement is key here, and likewise, in order to find understanding in relations, we have to really speculate as to what makes a person tick. Analyzing a relation is more complex than a controlled environment, so in order to do this, you must see more than what they are and why they are that way, but really who they are. The only way to understand this comprehensively is by "walking a mile in their shoes," which will require a level of care. Caring is a two-way street, meaning that you both have mutual interest in the outcome of the matter at hand, and through this process you will have discovered engagement. When you stand on this common ground, you will begin to see each other's point of view and will be able to more easily share and balance one another. Although the coming of media in our time has enabled various two-way interactions that serve as positive facilitation for this process, it also serves various one-way practices of new and old. Among others, an example of this evolution can be seen in the media version of "catcalling," leaving an inappropriate comment or pickup line with little to no expectation of reciprocity. This type of behavior comes from a lack of consideration for how you would feel in the other person's position and a disregard for whether or not your actions are genuinely aimed at forming a relation. These media have also enabled various two-way interactions where we can share things together and see where another is coming from in their feelings. An example of this type of interaction can be seen in literal sharing, retweeting, and @'ing as to directly include others in the relativity of any given topic or matter at hand. Reciprocated stimulation is important as nothing worthwhile comes without exertion of energy, and with regard to a relationship, shared experience requires each person to contribute comparable energy to maintain balance. With regards to balance and perception, it is important to maintain a relational equilibrium between having love for yourself and showing love to others. This reciprocation of energy is also applicable when there is balance in the amount of love you show others and the amount they show you. This harmonious love is a catalyst for producing peace and prosperity in the world, and is what we all strive for at the end of the day. Obviously, this is easier said than done with the complexity of forces driving our connectivity and relations in our modern era, circling back to the need of having truth in the love you give, receive, and hold for yourself. It takes courage and confidence in order to separate and distinguish love as true, and although it may bring about unforeseen challenges, it will certainly build character. Embrace this mindset as to understand perception as others do, and you will be concurrently letting down walls to allow an easier perception of truth. This will inevitably make you stronger as you continue to learn more about human nature without letting it control your perception—the ups and downs that come with temporary and lifelong relationships. In the end, it is this balance that we need because we are ultimately a reflection of others, and vice versa. We see what we want to in others depending on what we feel about ourselves during that given time, whether it be an undiscovered potential or straightforward feeling. The importance here as a result of this balance is to withhold judgment and to be accepting of everyone, especially those you care for in leading by example so they will be able to reciprocate the same for you.

Balance is of utmost importance for perception in order to maintain relational equilibrium, especially between engagement over media and physical presence. This concept is integral in proving that these media systems can either make or break your relations depending on the level of discipline you maintain. There are certain advantages the natural forms of engagement have over media forms, and vice versa, so it is your responsibility to utilize them properly in a manner that serves your individual desires. As with anything in life, moderation is the key to attaining sustainability, so as far as relational application goes we need to keep in mind that too much or too little of either media or natural engagement can be detrimental on a case by case basis. As an average, our millennial generation tend to overemphasize and overutilize the resources we have inherited with media expansion, so it is imperative we maintain a working proficiency of natural forms of engagement as well. As far as relational input goes, there will never be an equal amount of stimulus over technology and media as there is in the natural world. This is mainly due to the comprehensive nature of subconscious signals and in-depth intimate behaviors we can better send and receive through a natural environment. The unparalleled power of engagement through a natural setting is something that needs not only to be understood, but also practiced by this generation. Again, this is not to say that media engagement is inherently good or bad, rather that this generation tends to use it more than natural engagement, and we have a need to establish balance. Media capabilities have given way to many enhanced forms of engagement that could never be possible in a natural environment, such as communication from around the world. However, there are subtle nuances of a natural environment that are vital to relations, and the same really can't be said about these media, as the entire history of humankind before 100 years ago was able to relate with virtually no technology. This being said, yesterday is all but history, and it is only human nature that we change with the times, so it won't hurt to use our newfound capabilities with an appropriate level of balance. Once we have established an equilibrium between these types of engagement, we can focus on features of relational micro-interpersonal perception. A point to make about relational perception and equilibrium of love is that the same concept of importance in differentiation applies. Love is often paradoxical while still being balanced in these opposites very much like the yin and yang. Love can be make you feel inadequate and complete, ambiguous and certain, or even humbled and empowered all at the same time. There are truly countless stages of the relational process, so you will be successful by maintaining confidence in your personal perceptive judgment and ability to maintain balance within your own psyche. This will promote balance and stability within your loved ones and your relations as a whole. Alongside the idea of balance, you must learn to facilitate the transference of feelings as one person giving, the other receiving, and reverse order as necessary for feelings to be equally shared. This sense of shared feeling is the substance behind the clout of selflessness, in that the feeling together is worth more than what you could feel for yourself. This selflessness is a core concept of chivalry that empowers others and incentivizes their motivation to empower you and strengthen the bond of relation through unconditional love that perpetuates itself throughout the world.

The Millennial World

It is undeniable that the inheritance of technology has facilitated a new age of connectivity, having incidental effects on our societal and relational behavior. Side effects certainly extend beyond the millennial generation, but it seems that we have been most impacted by media. It can often be useful to look at some of the general symptoms, thereby setting a foundation to develop a cure for some if not all of these side effects. For starters, we prefer "Netflix and chill" over movie theater dates, we find it easier to have drunk texts than sober small talk, and we keep up with celebrities more than family and friends. Among other things, the effects of media we are so comfortable with has facilitated several other major influences on daily life and the relational principles we live by. Our comfortable and laissez-faire attitude is the cause of how heavily these influences weigh on our lives, to the extent that we have become desensitized to its effects on societal norms and relational basics that usually come second nature. An example of this concept can be seen in one of the most essential relational principles at stake, being the idea of privacy, or over media a lack thereof. By nature, relationships are meant to be private, especially in an intimate regard, and this is why monogamy is generally more practiced than open relations. Many aspects of a relation actually depend on exclusivity and parts being "for your eyes only," but it seems that we are now disclosing our personal life, and even those parts, as if it were now beneficial. Have we forgotten our fifth amendment and the benefits that come with the right to remain silent? It is often unwise and unhealthy to broadcast every development in your relationship because it takes away from the special experience that you share. Along with this new influence, there exists an age-old violation of privacy which has evolved with connectivity and media. This encompasses the other side of the spectrum of forfeiting your own privacy into the realm of violating another's privacy, referred to as cyberstalking. Never has there been such a normalization of this extremely flagrant concept. People are susceptible to the concept of stalking over the internet due to the ease of one-sided interaction and comfort that comes with the prospect of nonreciprocation without the other's knowledge. As previously mentioned, this one-sided engagement is a fundamental threat to our relational principles, and it can also be a threat to societal norms. One of these one-sided interactions is brought to you by obsession with celebrity and the idea of popularity over these media. People nowadays are literally buying followers to try to impress people they don't even know, or worse than money, spending their lives in search of 15 minutes of fame. The current state of infatuation with celebrity brought by the millennium has put more focus than ever on an illusion of worth through scale of measurable status and materiality, which is obviously unhealthy for a society as a whole. In regard to who you are to your community and what your relations mean to you, try to disregard the status quo and decipher value for the things life has to offer on your own scale of worth. This is a testament to the principle that it is not healthy to compare your subjective attributes, especially in terms of quantifiable measurement in this unreputable endorsement of the ego.

The perpetuated media outlook of our relational affairs often attributes value on a comparative scale, but it's apt to base this value on a scale of subjectivity. We see an effect of this in how this media enables our addictions to ego and self-righteous promotion, instead of an honest self-realization in one another. Aristotle advocated a theory of this in the ways of democracy, arising out of the notion that those who are equal in any respect are equal in all respects, and because we are equally free, we claim to be absolutely equal. Herein, we see the real purpose of media, in that it is not made for the facilitation of speculative judgment upon one another, rather to discover and elevate one another for our differences and similarities. This is what makes each relation valuable in its own respect because we have qualities that cannot be measured or traded, and this aspect is often what these media lack in portraying. Incipient media tendencies seem to mainly promote objectified mentalities, which are perpetuated by speculative analysis. However, this media can also serve as platforms to highlight the subjective beauty of our human nature and relations. In this concurrence, it is discovered that this is not the cause of the media, but rather our imposition of manners or "netiquette" we choose to abide by when deliberating our actions over them. This shows how the media standards of relational behavior are potentially obsolete outside the realm of technolcgy mainly due to the fact that vital organic elements of communication are lost. Appreciating each other for things that are not prefabricated through media will prevent misappropriating objectified media values in our everyday lives. These standards are linear with the industrialization of widespread connectivity and automated systems we should really only confide in beyond serving relational facilitation. This mentality comes down to a hierarchy of quantity over quality on a scale of relational function with media. Again, it is not to say that social media forces you to see this way, but it certainly is an enabler, added to the fact that they are currently pushing harder than ever using every possible outlet to affect how you use them to see quantifiable stats. Systems such as those being put into place in line with high traffic connectivity tend to perpetuate values of double standards for economic stimulus, so falling into this particular habitualization is where they objectively win and you subjectively lose. A perfect example of this is through industries that are involved with image, and how photoshopping in media form can directly transfer to the values we adhere to in person. These values are what the double standards imposed upon our society have created, much like how women are scrutinized for having both too much and too little makeup, given unrealistic Barbie expectations only to be condemned for plastic surgery—all to serve their manufactured social constructs. In a relational context, we are under pressure to have objective "relationship goals," but these standards are often contradictory to relational success. The point being here is simply to do it for you, not because you think someone else wants you to, especially if that someone is a systematic machine with no genuine stake in your true happiness. Love is truly the most subjective form of art there is—there should be no comparative scale of value imposed on you or your relations. This media is not meant to serve as a platform for judgment or criticism; it is there to provide encouragement of creativity and entertainment.

People tend to misconstrue the value to be attained from that which is popularized by media, but the majority of this utility is simply amusement. That is, most of the relevance in what we have to gain from media is not to be taken so critically, but to be entertained. A perfect demonstration of this concept is seen in reality TV and how 99 percent of the time, it is not real life or even based on something that should be considered a real-life situation. This concept has even spread to more critical matters, and we are beginning to see "fake news." Media has only been around for a couple decades, yet an even more recent trend which demonstrates this point is in the idea of memes. A majority of the media from which we attain information is not exactly aimed at building character and instilling real-life morals, or especially for the sake of this book, chivalry. This is surely not a blanket statement about the internet, as information technology is a powerful tool in reception of knowledge; however, for the purpose of relational skills, there are much better tools elsewhere. With respect to relations, we need to really speculate if what we are practicing over these media is aimed at building character, or if it is simply whimsical fun. Obviously, there is a middle ground of optimal utility in personal preference, but for the most part, it leans toward the latter of entertainment and should not be misconstrued otherwise. This concept is applicable to affecting the range from the microscale in attitudes and day-to-day behavior to the macroscale of communal and societal environment for serving relational sustainability and prosperity. Social media and other connectivity brought through the coming of the twenty-first century has become a significant part of our lives, but we must sort our priorities in which part of our image matters the most. A virtual profile giving representation to ourselves is nothing to be condemned, but it is important to remember where the distinction lies between relevance of that and real life, especially with relations. This is not to say that our connectivity is useless in relations as there are countless abilities that hold value to their core; they just tend to be more so directly involved with objective purposes, such as establishing a professional portfolio on LinkedIn. There is a time and place for all things, and oftentimes the place for personal relational productivity is beyond the ingenuity of media accessibility. You wouldn't seek to propose marriage over a text message, and likewise you would also not seek to make a grandiose occasion out of tagging a friend in a meme. There is a middle ground where we must find balance between technological and anthropological tendencies in order to optimally conduct our relations. Speculation of the millennial timeline shows how these media evolved side by side with our maturity and how this conditioning impacted our societal and relational dispositions. From the early stages of millennial childhood development, we were matched with systems that not even our parents knew how to operate or what they involved. An observation of this brief time period along the millennium shows how we have been conditioned to place excessive merit in the relational standards set by these media, which was not exactly the trade school of chivalry. Social media will continue to keep up with tech advancements in order to meet demand for networking adaptations, so it is imperative that we learn from where it all started as to not repeat this misplaced infatuation. If you take a step back from conditioned mentalities of millennial milieu and gauge the scope of our investment in these media, you will find that throughout history, we were able to manage relationships without these things all the same, if not better. There was a time before the millennium when we celebrated principles other than popularity and celebrity, and making use of this perception will help clarify the importance of these media ideologies versus those held in times of chivalry's debut.

The Millennial Gentleman

Basic speculation of dispositional bias about chivalry reveal characteristics which are exclusive to the male gender or gentleman, but this is a misconception. Even beyond the field of relations, the image we have revered in chivalry is both literally and figuratively embodied by the knight in shining armor; but I must reiterate, even in those times, one of the most influential figures to adopt these principles arose from a woman, Joan of Arc. An amazing feat of this millennium is in collective efforts to achieve equal rights and treatment, so now is the time to push our progress and include the implications this can have on our relations. In our modern world, we have so much potential in our capabilities, so it would be a waste to squander them on things that don't truly represent progress for humanity and civilization. Of course, completely different circumstances surrounded the times of the Middle Ages and the millennium; however, the basis of human conscience remains and we can learn from their codes. Not only did knights establish a code to serve the greater good of the world, but they seemed to effectively master relations on a microlevel as well. An adept skill in both societal dealings and personal relations inadvertently led to the idea of chivalry, which is cyclical in that it again serves you and the greater good of humanity. This brings us back to the base attributes both ladies and gentlemen possess that are encompassed by five noble vestiges: integrity, intention, intelligence, intuition, and initiative. Suffice it to say, all of these features are clearly transcendent to ethnicity, sex, religion, or any other currently perceived limiting factor, so chivalry should be practiced by everyone everywhere. The final all-encompassing attribute of chivalry is honor, which shares an inverse relation with all five traits as a proper application of any one or combination of the five can come as a result of, and result in, honor. This is the endgame of chivalry within these base attributes, and defines value over any reputable characteristic as it is true character in your moral strength. Knights abided by a code of honor because it was their lives at stake to the edge of a sword. Today, we do not adhere to such drastic measurements, but this does not have to be an excuse to let these standards escape us. Although we no longer practice medieval contests, a parallel in today's society is within the honor and respect to be attained by sports, and by extension sportsmanship. John Wooden said, *"Sports do not build character, they reveal it."* Dedication through blood, sweat, and tears is the foundation for honor in devoting your life to a level of discipline that will prove to translate in your character. There is no wrong in honor; it serves as great of a good as possible in its purest form, and the lack of honor is the cause of all evils. In a court of law, honorable is the title given to a judge, and we refer to them as your honor. This is a perfect parallel in that knights were held in a high position of authority with interpretation and enforcement of the law of the land and upholding justice. Today, we have laws based on communal ethics; however, back in the days of chivalry, the code of honor applied more to moral obligation. This held each person responsible for what they knew was right deep in their heart rather than government statute, and few were to be deemed worthy of leadership and knighted. It is honor that upholds true justice in the ultimate form of love and strength, putting a higher cause before your own.

As a constituent upon the code of honor in upholding chivalry, there is a quality that bears utmost importance in the world today, and this is humility. A central theme pertaining to the surrounding content is the idea of balance, and in respect to chivalry the adjective "well-rounded" is a perfect description of what it means to be humble in chivalry. It is not enough to have achieved a level of excellence in relational practice to be considered chivalrous. It is the realization of equality in potential for your fellow brothers and sisters that leads to having chivalry. Having conviction to do right by your people and your loved ones requires the humility to understand that every life holds intrinsic value. It seems that with the coming of our media, the idea of humility lost traction and we began to showboat and criticize more. A person's acts need only to be noticed by loved ones, and even when you receive external recognition, it is not appropriate to bathe in glory for purpose of feeding one's ego. A person of chivalry holds a diplomatic reputation by maintaining manners that call attention to themselves. Society has recently developed a propensity for competition of who can prove their worth through media, but the folly here is that this inflation of ego devalues the standing of our character through overcompensation. Media will condition you to think that it is important to show off whatever you control, but it is the search for control that leads to hubris and counters humility. This deprivation of humility in hubris is when you begin to think of yourself as superior to your fellow brothers and sisters, and it can be dangerous. With current media, it is easy to put your two cents in and potentially be hurtful or offensive. But who are we to pass judgment upon those we don't truly know who have done nothing to hurt us? Speculation of recent history before this global connectivity shows that apparently human beings are capable of despicable atrocities, and we have a very long way to go before we reach a truly globalized state of love. Even today we see racism, sexism, religious discrimination, and other general discordance in toleration of one another. It is imperative we not repeat history and reserve judgment upon our fellow brothers and sisters. We need to learn that we are all equal, and these media systems should be used to bring us closer to this realization in progress of relational connectivity and civilization. Not only does humility entail a resistance to flattery and overly boisterous display of aptitude, but it is in knowing when your cup has overrunneth. Knowing when enough is enough bears testimony to the fallacy of absolutes, in that absolute power corrupts absolutely. It is the abuse of power as a direct source of corruption that led to a demise of the feudal system and indirectly knighthood and chivalry during the Middle Ages. As a result of kings, lords, and dukes losing control over the lower class with emerging opportunities brought by mercantilism, a new world was born from the ashes. This meant that people would have newfound outlets in which to profit and carve their path without subjugation or governing limits. Our millennium is comparable with this shift in collective societal mentality and rush of connectivity enabling countless ways of life, especially with regards to relationships. So, it is imperative that we use these capabilities for the greater good when presented with the opportunity to abuse power. This sounds like a great undertaking by effect of implications, but it is really a minor adjustment to moral integrity. In retrospect, it is with luck that we find ourselves in this position of setting precedent in having such powerful technology, and although this may be a difficult position to be in, we have the most potential for progress. These systems should be used to empower us in achieving things once thought impossible, so we must not let them control our level of egocentricity, and rather use them in facilitation of chivalry.

With the abundance of capabilities at our disposal, we have become prone to being insatiable—have strength to be altruistic and pure of heart in benevolence. This is the single most important principle of chivalry, in that love is strength. Something that we all have by right of birth is the capacity to build and empower not only ourselves, but even more importantly each other. With regards to relationships in this connectivity, there is a necessary set of skills required to be affluent and prolific, but no one is born with the aforementioned traits—they are earned. Like earning anything, you must pass trials and tribulations to reach success, and this will often push your limits and willpower. The single most difficult thing to do is to show love in the face of ignorance and hardship, but chivalry will persevere in unyielding determination. We must learn to give others the chance we would want, regardless of how challenging it may be. Aristotle's teachings display this concept of love in how "love is composed of a single soul inhabiting two bodies."[1] which is recognition of oneness among humankind. This reinforces the concept of why you should have empathy to the point that you perceive your actions to affect another person as if it were yourself. Aristotle made sure to instill this upon his favorite student, Alexander the Great, who displayed great attributes of chivalry in this regard. He showed respect to other nations in resisting tyranny and was not great because of the fact that he conquered, but more because he united. After winning his battles and wars, he would offer kings the option to continue to rule by consent to joining his empire. Alexander said, "there is nothing impossible to him who will try,"[2] which is significant to the millennial era of abundance and ease of connectivity in our obligation to at least try to love. Alexander's life was prolific in that his legacy was a world more connected, because he believed in one nation for all. His was an empire of the mind, and its greatest tragedy was not in the loss of his lands, but more that the libraries of Alexandria were burned. This being said, what is to give light must endure burning. This is the basis of chivalry: It is the sacrifices we face with a smile because of the principle in undeterred ambition. Yes, to add to this we live in a time of turbulence and doubt in our relational skills, but it is a necessary step in creating prosperity for the future. In this time of commotion, we must stand out to create balance, not letting these things control us nor trying to control them, rather living together in harmony. The general message here with regards to our connectivity is not to provoke an epiphany where you delete your accounts and turn on a figuratively permanent airplane mode to be one with immediate surroundings. I am not suggesting you go off the grid; I only suggest that you redefine the grid to fit your true desires and find the possibility of giving love for everyone. There are many facets of character to embody a lifestyle of diverse composure in whatever ways bring about meaning to your ambition. The point here is not only to be efficient and effective in what you seek to become, but in the things that are more challenging as well. This includes patience when there is will to be hasty, prudence when there is will to be explicit, perseverance when there is will to give up. This is where you will gain the most utility from chivalry, being the noble cause of serving your community and contribution to the greater good of the world and to future generations. Showing this selflessness in paying it forward has a much larger impact than you can imagine, which is why it is most important to have balance in being able to control your inputs as to produce an outcome for prolific aspirations.

The Millennial Relations

In terms of relations, what people want can be basic and complex at the same time, like the paradoxical idea that the little things make the biggest difference. In a philosophical view of relations, that which leads to prosperity or failure is a ratio of how those involved value or devalue one another for the things they do. Aristotle said,

> Excellence is an art won by training and habituation. We do not act rightly because we have virtue or excellence, but we rather have those because we have acted rightly. We are what we repeatedly do. Excellence, then, is not an act but a habit.[3]

Applying this to relational value, it would seem best to have constant discipline in the ways of habitual attention to detail, and this is something our current media offers in abundance. Through connectivity, we are able to keep in touch and relay numerous smaller signals of care as opposed to relying on the few and far between grandiose occasions of extreme importance. In this millennium, it is unavoidable to cross paths with the bearing that media has had on our daily agenda and micro behavioral tendencies, so to deny its existence completely would deprive you of the little things it has to offer and ultimately stifle the potential of your relations. Keeping up with the times comes highly recommended for success of relations to our generation in order to provide the little things that most of us want and expect. Technology at its finest is not meant to replace portions of our lives, but rather to enhance and make them more efficient. This being said, to exclusively maintain a relationship over social media and technological communication is not sufficient. There is undoubtedly a near infinite amount of "little things" at our disposal in the real-world face-to-face realm of relations. This is the most important contribution to our relations, because this is where the majority of our true energy can be transferred intuitively. The complexity of what little things and people want the truth, and there is an unmatched conveyance of this in the little things people do subconsciously in the form of body language, voice intonation, and other natural communication. Communications obviously include a degree of physical interaction, and with respect to relationships, one of the most basic and complex forms of this is seen in the most important of areas—sex. Of course, recent advancements have made this act more accessible, such as the invention of the phone with phone sex and sexting; however, the real thing is irreplaceable. It should certainly not be replaced with mecha-sex robots. Natural sex is just one of many examples in how an action so little can be so greatly productive and even solve the biggest of dilemmas.

Aside from obvious occasions when it is called for such as birthdays, anniversaries, and so on, this can be helpful in day-to-day solutions and contextual random acts of kindness. If you can't sleep, you missed exercise, you're stressed out, or you're just bored, this can be the answer to your relational turbulence that our media simply can't offer. The point here is not just about benefits of a natural interaction, but rather in the fact that although media can be a great outlet for the little things, it will always fall short of the necessity for real-life relational interaction and the little things therein. In our relations, we have the power of the present, to go out of our way and show another person that they matter, no matter how big or small of a gesture it may be. You will tend to find that these little things are what people really care about at the end of the day, and this is where chivalry will come to serve you best. The more significant aspects of a relation come as a result of the work we are willing to constantly devote to the little things. As well as the benefits of smaller effects leading to larger purpose, there is a much deeper variety of little things you can do to choose from. Modern technology in media and connectivity offer a near infinite amount of various tools to facilitate our relations. These media have the potential to be a prolific catalyst toward transcending human limitations in a phylogenetic relationship between two or more groups of people resulting in a resemblance in general plan or structure, and in the essential structural parts. Recognition of the almost infinite little things that all people share in common will lead to the creation of the widespread mentality that we have a fundamental empathy in chivalry.

One of the most powerful attributes of relations is that it is unlimited to any form of acceptance; there is room for strength in love everywhere and for everyone. There are over 1,000 languages in the world, and love is able to speak them all. Where there is a will there is a way; we just need to have strength to love the way we were made to. In our millennium, we tend to practice selective love based on what media suggests is currently acceptable. Just because you did not give or receive a certain number of likes on different statuses or pictures, does not mean you are not capable of being, or are already abundantly loved. This comes back to the breadth vs. depth of love comparison. In our current society, we often judge and are judged for who we associate with, but we are all human beings and we are more alike than we are different. This same concept applies to a micro-singularity, and how we can always find something to love about anyone. In taking the good with the bad, you will avoid loving people on and off in pieces, or picking and choosing what you love about them. If you love someone, love everything about them and full heartedly. With regards to this concept, you must remember that everyone comes with a past, and you shouldn't focus too much on old news because people change. Within our age in modern connectivity, a large portion of people's lives is on display, so you must not dwell on the past; instead simply live in the present, and look to the future. Along the lines of overcoming people's baggage, it is important to have the wherewithal to see others without any biases of your own comparisons. One person's timeline is completely disproportionate to another's, and your own for that matter. Regardless of age, status, or any other notable merit, you must see that we all have our own path and although it is not predetermined, in the grand scheme of things, people are going to do what they want and these things perfectly align with what makes them who they are, not who you are. Part of accepting yourself is admitting that some things will never be perfect, but some people can be perfect for you. In order to find this, you must learn to accept those you love for their flaws, whether it be physical, mental, or spiritual. Some things can be changed and some things cannot, but the important part is that you love people for who they are as a whole and not concentrate on what you may find different in any particular aspect. Not only will it help you to practice tolerance in this way, but also in having control over the type of influence you have on others, and vice versa. In terms of a successful relationship, you may want to consider what type of environment you are surrounded by. To maintain strength in a relationship, make sure that whatever you do appeals to the mutual satisfaction of the both of you, instead of folding to the pressures of others or pressuring them. In this concept, we not only find strength in our personal relations, but on a global scale of civility among all humanity. In order to achieve this, we simply need to communicate to one another that we matter. All lives matter equally, which is something our media is beginning to shed light on, and in this unity we can find strength. Among all levels of biological species, we find that the most resilient and thriving are those who cooperate and coexist in supporting one another, so it only makes sense that as the most advanced species on this planet, we should be too. We are the most sophisticated life form on earth, so it is imperative that we maintain and harness these abilities to lift one another up in efforts to create a living space that we can share as neighbors in a globalized love.

Although media provides useful resources in the facilitation of our lives to a point, discretion ultimately falls on the people to decide what really matters. In line with everything subjective in life, you have complete and total power to determine how it affects you, so don't be dependent on the standards of others. In a relational respect, this is important to how you base your social preferences on intuitive instinct. Even with all of the amazing things media has brought us, in the end, all that will truly matter is what we have to offer in person. Some of these things can be cross-compatible, but oftentimes they will be restricted to their respective realms. It is important to have organic engagement, and although these media often lack the proficiency for this and may never reach a full capacity, they are advancing similarities in replication. An example of this concept can be seen in "read receipts" or the media's system for letting a sender and receiver know when one or the other has engaged in conversation. Technology will continue to grow and advance in this detail, so it is our responsibility to clearly understand balance of our lives involving the two in coexistence. This goes back to the concept of relational breadth vs. depth, and how it is usually best to have your relations involved with both technical media and real-world experience. This being said, it should be noted that these systems lack the capacity that humans inherently have for truth in undefined energy that is only understandable through intuitive real-world forms of interaction. The exact precision of this universal language in natural energy tells us the perfect answer through instinctive guesswork as the universe gives us confirmation. A serenity of balance is what we all seek, but the truth is that the scales of life are never stagnant. Life will have its ups and downs in certain areas, and although experience is constantly shifting different rates, that is not important as long as the energy is channeled toward making your life and relations succeed. In order to succeed in relations, you do not need to direct 100 percent of your total energy, as this is one-sided and would leave nothing for yourself. It is necessary to dedicate 100 percent of the energy any other person or people need; if this is more than what you have to offer, they are simply too needy. In this case, the significance of the relation is probably less than the case of those little things being majoritively involved with real-world interaction. It is also important to dedicate a relatively consistent amount of energy in maintaining what standards you have set, treating the 1,000th day like the 1st and so on. Despite what fairy tales may suggest, your knight in shining armor or fair maiden will not come riding into your life on a white stallion or wait in a high tower for you. You may also find that media plays a large role in the prosperity of your relations. The reality is that you must accept any forms of productivity when you apply the strength of available tools with harmony and clairvoyance. The takeaway with regards to the basis of relations is to find balance, catering to the common ground that you and your loved ones share, and not false perception of what the world including people you don't even know expect of you over media. Go after the person or people who make giving yourself, a gift in itself. Go after the person who will have you for nothing more or less than who you are, and accept them for the exact same in return. Devote your energy trying to prove your love to the people involved in facilitating it, and not trying to prove things to the people who aren't. Your time in this life is limited, so don't waste it living someone else's. The acronym YOLO[4] is in the *Oxford English Dictionary*, and the definition will show you why you should stop making your life about trying to impress other people. With everything going on in the world today, it is easy to become distracted and follow social trends, and although it is important to keep up with the times, you must form with them instead of letting them control you. As much as societal forces through media seek to help or hinder us, you are the omega in your dreams and fulfillment in what really matters.

The Message

As with anything subjective in life, there is not any single answer or designed procedure, but rather a variety of approaches in a comprehensive solution. Variety is the spice of life, and coinciding with applicable moderation, this allows you the freedom to enjoy everything that life has to offer in appropriate proportions. In a relational respect, this concept of multifaceted persona is noted in the idea of universality and the catch-22 that we are all of the same existential reality, but also individually

unique. This idea of universality helps decipher the enig-
ma of free will and how our individual choices impact
the internal and external aspects of our relations. In our
millennium, this idea holds major significance in how
we are designed to connect with one another, and in a
media sense, this idea of technological singularity. The
effects of this interaction extend beyond our relations
and immediate surroundings to echo in all directions for
the remainder of time. Due to our finite supply of time
and energy along the spectrum of eternity, there is an
urgency in self-realization and honing in on what will
complement the unique "it" factor of your being. Even
if "it" is currently a *"je ne sais quoi,"* you will still feel
something, and you must embrace and refine whatever
this is to reach positive realization. This will only come
by means of staying true to yourself, which is not something you will be able to solely accomplish in a limited
environment like some of these media, so again you must
find balance. This being said, although finding yourself
is more easily done without certain media, the purpose
of relations is to more easily find others, which is something our media provides in abundance. The small caveat herein is not forgetting who you are when navigating
the astronomical variety of ulterior persona among these
media outlets of connectivity. There is a certain scope of
recognition within who you are, who you were, and who
you want to be. In perceiving current relations with perspective to the past and regards for the future, we see that
now is a pivotal point in time with all that we are being
presented. It is this analysis in attempt to draw personal-
ized conclusions that will serve a greater purpose than

our own, eventually recognizing that they are one in the same. In life, we may reach many solutions surrounding the same conclusion simply because we continue to learn and see things from a different perspective. This idea can be seen in the utility of our media, and how it offers a globally cultured take on human life and relations in how we find ways to share information and help each other answer all of life's questions. To find answers, a question is not always entailed the binary yes/no response; in fact, relationally subjective questions depend on a much deeper inquiry and investigation of reasoning to find truth—a testament to the effectiveness of comprehensive solution. The most simple and effective way to find truth is through the unyielding test of time, as all will be revealed if you continue to press boundaries. Timing is everything; when opportunity meets preparation, the result is success, and "they all lived happily ever after." Having a digital world of instant connection at our fingertips can distract us from the perception of how precious time is simply because we are lost in it. Just because the pieces fit doesn't mean they go there— another testament to the importance of comprehensive solution. Likewise, just because your time can be spent doing something does not mean it should be done. Our media has amazing benefits to its credit, but it has a limit of utility in the ways of creative expression, which is something relations demand. The beauty in life truly comes from the signature strokes that make the pieces fall in a lane of their own. There is a reason handwritten letters are so much more intriguing than emails, even if it is the same exact verbiage. All things come in good time, so to use connectivity to expedite relations is a major disservice to yourself, as relations are much harder to repair than they are to form. When you are able to properly gauge the value of this time, you will have found sustainability in an environment productive for any successful relationship. At the end of the day, your best use of time will be in the moments you share with those you love, whether that be over media or in person is immaterial, as long as it is unconditional.

Unconditional love is the key to relational fulfillment due to the nature of reciprocity, especially over media engagement and the expectations of validation. Arguably one of the most important aspects of this concept is in the ability to find unconditional love for yourself, as this will serve as a foundation for your relations. We tend to hold back the true nature of our personality over media and real-world interaction because of a conditional love for ourselves and the prospect of somehow being vulnerable. This is usually more common among media because we receive different levels of validation in quantifiable measurements, but really the only validation you need is your own. A lack herein affects how we feel about ourselves, indirectly how others feel about us, and results in limiting the validity of relations. The level of validity that you hold your relations to should be something that is mutually felt in the sense of subjectivity in qualifiable measure, instead of objectivity in quantifiable measure. Whether engaging in person or over media, we have the ability of feeling and allowing others to feel our energy, and this transference is boundless. There are countless examples of media facilitating activism in the form of viral topics, so why not apply this concept to connection in unconditional love for the world to share. The feeling of connection can be felt face-to-face or miles away, so a conditional love for others in an exclusively specific time or place can result in a disconnect when these criteria are not met. To the respect of shared feelings, it is impossible to avoid truth; even when it is not known, it is felt. Thus, an unconditional love for others will serve your relations most effectively as it does not change throughout different circumstances. This is a defining characteristic of reciprocity, in that a transference of energy can be felt in the purest of forms through the idea that love is strength, hate is weakness, and the parties involved dictate relational stability through manifesting their true feelings. These media offer an outlet for this projection through the abundance of opportunity in the form of devices that provide optimal exposure to continuous updates on your personal life. Although we can be quick to engage in this activity, it is important to remember that this is a limited environment, so effects will be conditional and secondary to real-world engagement. Here, the importance of unconditional love from being so connected is seen in that you reap what you sow, wherein the effects of your actions extend beyond a foreseeable outcome, and will certainly come back to you in one way or another. People will crave connection in any form, which is why this media is so popular, but it is easy to get too involved and dependent on the different types of love this media can provide. Unconditional love comes with no strings attached, so it is beyond any number of likes and comments. These would be used in a case of conditional love based on having some sort of expected quantifiable validation, and include only parts of a message that do not capture the true essence of your expression or the response—it is lost in translation. Chivalry is not applicable to this type of love due to the basic principle of selflessness, and simply providing something for the merit of goodwill and inspiration without need for quantifiable validation. The purpose of unconditional love is in how we choose to reciprocate independent of others' disposition to giving anything back. Even being that actions of unconditional love need not to be returned, it does not matter because they serve you just as well. As melodramatic as it may sound, unconditional love is a gift better enjoyed when given, simply because it is a free investment in the best capital, and returns are unlimited.

In retrospect, a lot of the aforementioned content is centered around this millennium, but ideas of unconditional love and chivalry transcend time and space. There is a lot of speculation as to whether the idea of chivalry is currently at risk, and that a golden age of chivalry has come and gone, but our millennium is truly golden. We were amidst and most directly involved in these millennial phenomena, so we already know what has just been mentioned. For the time being, this book and the message to be taken from it is mostly in place for future generations, so they can better relate to us and themselves. When the time comes that we find ourselves elders, we must take this message as a case study of our younger selves to see how we can relate to the future and the generations that we produce. Contrary to popular belief, we are not products of our environment—we produce our environment. Technologic and systematic innovations had been made because there was a demand for them, and good or bad, we learn. At the end of the day, we are sentient beings, and these are only tools of our trade so we should see these things as stepping stones for future capabilities in our relations, and chivalry. We must not forget the importance of our basic essential being and trust of instinct and genuine experience. We must follow what we believe to be honorable and good based on conscience. It is apparent that people are capable of good and bad things, but to our core, we know what is right and wrong and we are inclined to do good to others because it is what we would want to be done to us. Do not be misled by synthetic or virtual representations of what people are telling you to believe based on their interests. It's all about your attitude, and at the end of the day, you need to have a positive outlook to keep your head up. With emphasis in connectivity media and information systems at our disposal, it is easy to treat engagements as disposable. Try to comprehend that we are a very small select few who are capable of such beautiful relations. These, and all other relations are priceless. There's more to life than temporary benefit and instant gratification, so concentrating on balance and sustainability is healthy for your relations. Whether it be sustaining our world or our relations, you must learn to take only what you need and to give back that which you can. Chivalry extends beyond relations; it is embedded in humanity that we should lift one another up when there is less cost to ourselves than would be for another. Chivalry is a set of good principles that remain constant, but as times change so does its utility. In our millennial age, we now have far more capabilities than previous generations, so it is now a matter of greatness. Your projection of energy and the people you positively affect along the way is your life's legacy, so greatness is the decision now. History rewards the bold, and in this millennial age, we have ease of opportunity to separate ourselves and emphasize the signature abilities that we have been blessed with, as they are truly intended for the world to see. You must be brave and go full heartedly into the unknown, accepting failure and hardship along the way to your true calling. It is the chance taken that is always the source of creation, whether you succeed or fail in your first attempt, there is always something to be gained. Chivalry is about understanding the treachery that lies ahead and staring it in the eye with composed demeanor and vindication. Use this ambition in your relations to treat people with an understanding that they too face challenges unknown to you, and share sources of strength and inspiration. If there is anything the World Wide Web has to show us, it is that we are truly connected in more ways than we can comprehend. With this in mind, it is up to you, the individual, to be a part of a collective movement. Chivalry has had peaks and valleys in the actions of humankind throughout history, and it is at this point in time that we must decide whether to build upon or destroy its code as an effect of our capabilities. It is our duty to utilize what we have been given in a responsible, positive, and loving manner. It is in this feeling of guidance that the mind follows the heart. When love is felt, there is no explanation necessary; with its absence, no explanation is possible. Love is the driving force that fuels chivalry, and is the strongest impetus of the human spirit. Once you discover why love is so important, you will value it over everything, and everyone you share this feeling with will value your love above anything else you could hope to offer.

Notes

Chapter 1: The Millennial Perception

1. Saylor Academy, "Rhetorical Appeals," 2012, https://www.saylor.org/site/wp-content/uploads/2012/12/Engl001-1.1.2_Rhetorical-Appeals.pdf.

2. Mind Tools, "VAK Learning Styles," https://www.mindtools.com/pages/article/vak-learning-styles.htm.

3. Lucid Smart Pill, "Your 5 Brainwaves: Delta, Theta, Alpha, Beta and Gamma," https://lucid.me/blog/5-brainwaves-delta-theta-alpha-beta-gamma/.

4. The Darwin Project, "Darwin's Unfolding Revolution," http://www.thedarwinproject.com/revolution/revolution.html.

Chapter 2: The Millennial World

1. Everyday Health, "What Is Adderall (Adderall XR)?" https://www.everydayhealth.com/drugs/adderall.

2. XML, "Nokia Unveils the World's First Media Phone for Internet Access," http://xml.coverpages.org/nokiaWAP9902.html.

3. Startup Bros, "Myspace–The Rise, Fall, and Rise Again?" https://startupbros.com/myspace-the-rise-fall-and-rise-again-infographic/.

4. Devon Glen, "The History of Social Media from 1978–2012," *Adweek*, February 16, 2012, http://www.adweek.com/digital/the-history-of-social-media-from-1978-2012-infographic/.

5. Alyson Shontell, "How Instagram Co-Founder Kevin Systrom Spent His Year after the $1 Billion Facebook Acquisition," *Business Insider*, May 9, 2013, http://www.businessinsider.com/its-been-1-year-since-facebook-bought-instagram-for-1-billion-heres-how-co-founder-kevin-systrom-spent-it-2013-5.

6. Natasha Zarinsky, "What the Hell Is Up with 'Bae'?" *Esquire*, July 25, 2014, http://www.esquire.com/lifestyle/news/a29423/where-did-bae-come-from/.

Chapter 3: The Millennial Gentlemen

1. Nova Online, "Song of Roland," http://novaonline.nvcc.edu/eli/evans/his101/notes/roland.html.

2. "The Canterbury Tales: General Prologue & Frame Story by Geoffrey Chaucer–The Knight," https://www.shmoop.com/canterbury-tales-prologue/the-knight.html.

3. Léon Gautier, *Chivalry: The Everyday Life of the Medieval Knight*, (G. Routledge and Sons, 1891). [English edition.]

4. *Star Wars II: Attack of the Clones*, directed by George Lucas (2002; San Francisco, CA: Lucasfilm), Film.

5. Francis Beaumont and John Fletcher, *The Woman Hater* (Michigan, USA: Gale Ecco, 2010). [Originally published in the Stationer's Register, 1607.]

6. Anon. (This quote is wrongly credited to Einstein in the general media.)

Chapter 4: The Millennial Relationship

1. Alfred Tennyson, "In Memoriam A. H. H. OBIIT MDCCCXXXIII: 27," https://www.poetryfoundation.org/poems/45336/in-memoriam-a-h-h-obiit-mdcccxxxiii-27.

Chapter 5: The Millennial Message

1. Philosiblog, "Love Is Composed of a Single Soul Inhabiting Two Bodies," October 29, 2012, http://philosiblog.com/2012/10/29/love-is-composed-of-a-single-soul-inhabiting-two-bodies/.

2. Sanjana Ray, "'There Is Nothing Impossible to Him Who Will Try'–Leadership lessons from Alexander the Great," *Yourstory*, https://yourstory.com/2016/09/there-is-nothing-impossible-to-him-who-will-try-leadership-lessons-from-alexander-the-great/.

3. Think Exist, "Aristotle Quotes," http://thinkexist.com/quotation/excellence_is_an_art_won_by_training_and/10320.html.

4. BBC News, "Moobs and YOLO among New Words in Oxford English Dictionary," September 12, 2016, http://www.bbc.com/news/uk-37336564.